Marx Joyce
Abbott Hardy Machiavelli Chesterton Austen
Defoe Melville Montaigne Cooper Hugo
Haggard Emerson Eliot Grimm
Stoker Christie Molière
Wilde Carroll Maupassant Byron Schiller
Garnett Fitzgerald Engels
Goethe Einstein Hawthorne Smith Kafka
Cotton Dostoyevsky Hall
Baum Kipling Doyle Willis
Dumas Henry Nietzsche
Leslie Flaubert Turgenev Balzac
Stockton Vatsyayana Crane
Burroughs Verne
Curtis Tocqueville Vinci
Homer Widger Tolstoy Gogol Busch
Darwin Thoreau Whitman
Potter Freud Zola Twain Scott Plato
Kant Jowett Lawrence Harte
Stevenson Dickens Burton Hesse
Andersen
London Descartes Voltaire
Poe Aristotle Wells Cervantes Cooke
Hale James Hastings
Bunner Shakespeare
Richter Chambers Irving
Doré da Benedict Alcott
Dante Shaw Pushkin
Swift Chekhov
Wodehouse Newton

tredition®

tredition was established in 2006 by Sandra Latusseck and Soenke Schulz. Based in Hamburg, Germany, tredition offers publishing solutions to authors and publishing houses, combined with worldwide distribution of printed and digital book content. tredition is uniquely positioned to enable authors and publishing houses to create books on their own terms and without conventional manufacturing risks.

For more information please visit: www.tredition.com

TREDITION CLASSICS

This book is part of the TREDITION CLASSICS series. The creators of this series are united by passion for literature and driven by the intention of making all public domain books available in printed format again - worldwide. Most TREDITION CLASSICS titles have been out of print and off the bookstore shelves for decades. At tredition we believe that a great book never goes out of style and that its value is eternal. Several mostly non-profit literature projects provide content to tredition. To support their good work, tredition donates a portion of the proceeds from each sold copy. As a reader of a TREDITION CLASSICS book, you support our mission to save many of the amazing works of world literature from oblivion. See all available books at www.tredition.com.

 Project Gutenberg

The content for this book has been graciously provided by Project Gutenberg. Project Gutenberg is a non-profit organization founded by Michael Hart in 1971 at the University of Illinois. The mission of Project Gutenberg is simple: To encourage the creation and distribution of eBooks. Project Gutenberg is the first and largest collection of public domain eBooks.

Bird Stories from Burroughs
Sketches of Bird Life Taken from the Works of John Burroughs

John Burroughs

Imprint

This book is part of TREDITION CLASSICS

Author: John Burroughs
Cover design: Buchgut, Berlin – Germany

Publisher: tredition GmbH, Hamburg - Germany
ISBN: 978-3-8472-1677-3

www.tredition.com
www.tredition.de

Copyright:
The content of this book is sourced from the public domain.

The intention of the TREDITION CLASSICS series is to make world literature in the public domain available in printed format. Literary enthusiasts and organizations, such as Project Gutenberg, worldwide have scanned and digitally edited the original texts. tredition has subsequently formatted and redesigned the content into a modern reading layout. Therefore, we cannot guarantee the exact reproduction of the original format of a particular historic edition. Please also note that no modifications have been made to the spelling, therefore it may differ from the orthography used today.

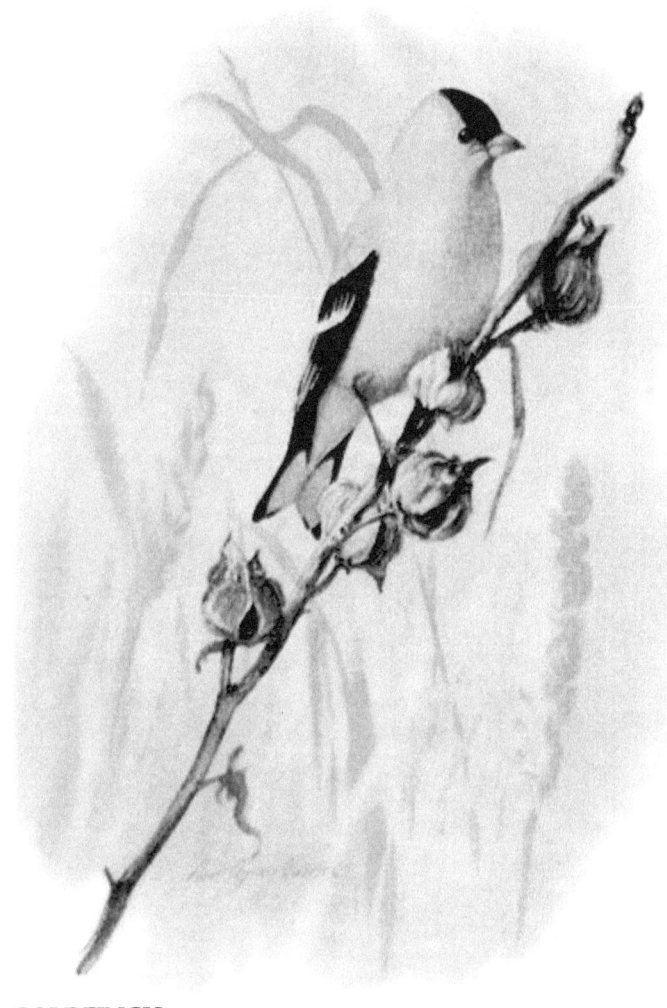

GOLDFINCH

BIRD STORIES FROM BURROUGHS

SKETCHES OF BIRD LIFE TAKEN FROM THE WORKS OF
JOHN BURROUGHS
WITH ILLUSTRATIONS
BY LOUIS AGASSIZ FUERTES

BOSTON NEW YORK CHICAGO
HOUGHTON MIFFLIN COMPANY
The Riverside Press Cambridge
COPYRIGHT, 1871, 1875, 1876, 1877, 1879, 1881, 1886, 1894, 1899, 1903, 1904,
1905, 1906, 1907, 1908, 1909, BY JOHN BURROUGHS

COPYRIGHT, 1911, BY HOUGHTON MIFFLIN COMPANY
Transcriber's Note: Hyphenation has been standardised. Minor typographical errors have been corrected without note.

PUBLISHERS' NOTE

John Burroughs's first book, "Wake-Robin," contained a chapter entitled "The Invitation." It was an invitation to the study of birds. He has reiterated it, implicitly if not explicitly, in most of the books he has published since then, and many of his readers have joyfully accepted it. Indeed, such an invitation from Mr. Burroughs is the best possible introduction to the birds of our Northeastern States, and it is likewise an introduction to some very good reading. To convey this invitation to a wider circle of young readers the most interesting bird stories in Mr. Burroughs's books have been gathered into a single volume. A chapter is given to each species of bird, and the chapters are arranged in a sort of chronological order, according to the time of the bird's arrival in the spring, the nesting time, or the season when for some other reason the species is particularly conspicuous. In taking the stories out of their original setting a few slight verbal alterations have been necessary here and there, but these have been made either by Mr. Burroughs himself or with his approval.

[v]

BIRD STORIES FROM BURROUGHS

THE BLUEBIRD

It is sure to be a bright March morning when you first hear the bluebird's note; and it is as if the milder influences up above had found a voice and let a word fall upon your ear, so tender is it and so prophetic, a hope tinged with a regret.

There never was a happier or more devoted husband than the male bluebird. He is the gay champion and escort of the female at all times, and while she is sitting he feeds her regularly. It is very pretty to watch them building their nest. The male is very active in hunting out a place and exploring the boxes and cavities, but seems to have no choice in the matter and is anxious only to please and encourage his mate, who has the practical turn and knows what will do and what will not. After she has suited herself he applauds her immensely, and away the two go in quest of material for the nest, the male acting as guard and flying above and in advance [2] of the female. She brings all the material and does all the work of building, he looking on and encouraging her with gesture and song. He acts also as inspector of her work, but I fear is a very partial one. She enters the nest with her bit of dry grass or straw, and, having adjusted it to her notion, withdraws and waits near by while he goes in and looks it over. On coming out he exclaims very plainly, "Excellent! excellent!" and away the two go again for more material.

I was much amused one summer day in seeing a bluebird feeding her young one in the shaded street of a large town. She had captured a cicada or harvest-fly, and, after bruising it awhile on the ground, flew with it to a tree and placed it in the beak of the young bird. It was a large morsel, and the mother seemed to have doubts of her chick's ability to dispose of it, for she stood near and watched its efforts with great solicitude. The young bird struggled valiantly with the cicada, but made no headway in swallowing it, when the mother took it from him and flew to the sidewalk, and proceeded to break and bruise it more thoroughly. Then she again placed it in his beak, and seemed to say, "There, try it now," and sympathized so thoroughly with his efforts that she repeated many of his motions and contortions. But the great fly was unyielding, and, [3] indeed, seemed ridiculously disproportioned to the beak that held it. The young bird fluttered and fluttered, and screamed, "I'm stuck, I'm stuck!" till the anxious parent again seized the morsel and carried it to an iron railing, where she came down upon it for the space of a minute with all the force and momentum her beak could command. Then she offered it to her young a third time, but with the same result as before, except that this time the bird dropped it; but she reached the ground as soon as the cicada did, and taking it in her beak flew a little distance to a high board fence, where she sat motionless for some moments. While pondering the problem how that fly should be broken, the male bluebird approached her, and said very plainly, and I thought rather curtly, "Give me that bug," but she quickly resented his interference and flew farther away, where she sat apparently quite discouraged when I last saw her.

One day in early May, Ted and I made an expedition to the Shattega, a still, dark, deep stream that loiters silently through the woods not far from my cabin. As we paddled along, we were on the alert for any bit of wild life of bird or beast that might turn up.

There were so many abandoned woodpecker [4] chambers in the small dead trees as we went along that I determined to secure the section of a tree containing a good one to take home and put up for the bluebirds. "Why don't the bluebirds occupy them here?" inquired Ted. "Oh," I replied, "bluebirds do not come so far into the woods as this. They prefer nesting-places in the open, and near human habitations." After carefully scrutinizing several of the trees, we at last saw one that seemed to fill the bill. It was a small dead tree-trunk seven or eight inches in diameter, that leaned out over the water, and from which the top had been broken. The hole, round and firm, was ten or twelve feet above us. After considerable effort I succeeded in breaking the stub off near the ground, and brought it down into the boat. "Just the thing," I said; "surely the bluebirds will prefer this to an artificial box." But, lo and behold, it already had bluebirds in it! We had not heard a sound or seen a feather till the trunk was in our hands, when, on peering into the cavity, we discovered two young bluebirds about half grown. This was a predicament indeed!

Well, the only thing we could do was to stand the tree-trunk up again as well as we could, and as near as we could to where it had stood before. This was no easy thing. But after a time we had [5] it fairly well replaced, one end standing in the mud of the shallow water and the other resting against a tree. This left the hole to the nest about ten feet below and to one side of its former position. Just then we heard the voice of one of the parent birds, and we quickly paddled to the other side of the stream, fifty feet away, to watch her proceedings, saying to each other, "Too bad! too bad!" The mother bird had a large beetle in her beak. She alighted upon a limb a few feet above the former site of her nest, looked down upon us, uttered a note or two, and then dropped down confidently to the point in the vacant air where the entrance to her nest had been but a few moments before. Here she hovered on the wing a second or two, looking for something that was not there, and then returned to the perch she had just left, apparently not a little disturbed. She hammered the beetle rather excitedly upon the limb a few times, as if it were in some way at fault, then dropped down to try for her nest again. Only vacant air there! She hovers and hovers, her blue wings flickering in the checkered light; surely that precious hole *must* be there; but no, again she is baffled, and again she returns to her perch, and mauls the poor beetle till it must be reduced to a pulp. Then she makes a third attempt, then a fourth, [6] and a fifth, and a sixth, till she becomes very much excited. "What could have happened? am I dreaming? has that beetle hoodooed me?" she seems to say, and in her dismay she lets the bug drop, and looks bewilderedly about her. Then she flies away through the woods, calling. "Going for her mate," I said to Ted. "She is in deep trouble, and she wants sympathy and help."

In a few minutes we heard her mate answer, and presently the two birds came hurrying to the spot, both with loaded beaks. They perched upon the familiar limb above the site of the nest, and the mate seemed to say, "My dear, what has happened to you? I can find that nest." And he dived down, and brought up in the empty air just as the mother had done. How he winnowed it with his eager wings! how he seemed to bear on to that blank space! His mate sat regarding him intently, confident, I think, that he would find the clew. But he did not. Baffled and excited, he returned to the perch beside her. Then she tried again, then he rushed down once more, then they both assaulted the place, but it would not give up its secret. They talked, they encouraged each other, and they kept up the search, now one, now the other, now both together. Sometimes they dropped down to within a few feet of the entrance to the nest, and we thought [7] they would surely find it. No, their minds and eyes were intent only upon that square foot of space where the nest had been. Soon they withdrew to a large limb many feet higher up, and seemed to say to themselves, "Well, it is not there, but it must be here somewhere; let us look about." A few minutes elapsed, when we saw the mother bird spring from her perch and go straight as an arrow to the nest. Her maternal eye had proved the quicker. She had found her young. Something like reason and common sense had come to her rescue; she had taken time to look about, and behold! there was that precious doorway. She thrust her head into it, then sent back a call to her mate, then went farther in, then withdrew. "Yes, it is true, they are here, they are here!" Then she went in again, gave them the food in her beak, and then gave place to her mate, who, after similar demonstrations of joy, also gave them his morsel.

Ted and I breathed freer. A burden had been taken from our minds and hearts, and we went cheerfully on our way. We had learned something, too; we had learned that when in the deep woods you think of bluebirds, bluebirds may be nearer you than you think.

One mid-April morning two pairs of bluebirds [8] were in very active and at times violent courtship about my grounds. I could not quite understand the meaning of all the fuss and flutter. Both birds of each pair were very demonstrative, but the female in each case the more so. She followed the male everywhere, lifting and twinkling her wings, and apparently seeking to win him by both word and gesture. If she was not telling him by that cheery, animated, confiding, softly endearing speech of hers, which she poured out incessantly, how much she loved him, what was she saying? She was constantly filled with a desire to perch upon the precise spot where he was sitting, and if he had not moved away I think she would have alighted upon his back. Now and then, when she flitted away from him, he followed her with like gestures and tones and demonstrations of affection, but never with quite the same ardor. The two pairs kept near each other, about the house, the bird-boxes, the trees, the posts and vines in the vineyard, filling the ear with their soft, insistent warbles, and the eye with their twinkling azure wings.

BLUEBIRD
Upper, male; lower, female

Was it this constant presence of rivals on both sides that so stimulated them and kept them up to such a pitch of courtship? Finally, after I had watched them over an hour, the birds began to come into collision. As they met in the vineyard, [9] the two males clinched and fell to the ground, lying there for a moment with wings sprawled out, like birds brought down by a gun. Then they separated, and each returned to his mate, warbling and twinkling his wings. Very soon the females clinched and fell to the ground and fought savagely, rolling over and over each other, clawing and tweaking and locking beaks and hanging on like bull terriers. They did this repeatedly; once one of the males dashed in and separated them, by giving one of the females a sharp tweak and blow. Then the males were at it again, their blue plumage mixing with the green grass and ruffled by the ruddy soil. What a soft, feathery, ineffectual battle it seemed in both cases! — no sound, no blood, no flying feathers, just a sudden mixing up and general disarray of blue wings and tails and ruddy breasts, there on the ground; assault but no visible wounds; thrust of beak and grip of claw, but no feather loosened and but little ruffling; long holding of one down by the other, but no cry of pain or fury. It was the kind of battle that one likes to witness. The birds usually locked beaks, and held their grip half a minute at a time. One of the females would always alight by the struggling males and lift her wings and utter her soft notes, but what she said — whether she was encouraging [10] one of the blue coats or berating the other, or imploring them both to desist, or egging them on — I could not tell. So far as I could understand her speech, it was the same that she had been uttering to her mate all the time.

When my bluebirds dashed at each other with beak and claw, their preliminary utterances had to my ears anything but a hostile sound. Indeed, for the bluebird to make a harsh, discordant sound seems out of the question. Once, when the two males lay upon the ground with outspread wings and locked beaks, a robin flew down by them and for a moment gazed intently at the blue splash upon the grass, and then went his way.

As the birds drifted about the grounds, first the males, then the females rolling on the grass or in the dust in fierce combat, and between times the members of each pair assuring each other of undying interest and attachment, I followed them, apparently quite unnoticed by them. Sometimes they would lie more than a minute upon the ground, each trying to keep his own or to break the other's hold. They seemed so oblivious of everything about them that I wondered if they might not at such times fall an easy prey to cats and hawks. Let me put their watchfulness to the test, I said. So, as the two males [11] clinched again and fell to the ground, I cautiously approached them, hat in hand. When ten feet away and unregarded, I made a sudden dash and covered them with my hat. The struggle continued for a few seconds under there, then all was still. Sudden darkness had fallen upon the field of battle. What did they think had happened? Presently their heads and wings began to brush the inside of my hat. Then all was still again. Then I spoke to them, called to them, exulted over them, but they betrayed no excitement or alarm. Occasionally a head or a body came in gentle contact with the top or the sides of my hat.

But the two females were evidently agitated by the sudden disappearance of their contending lovers, and began uttering their mournful alarm-note. After a minute or two I lifted one side of my hat and out darted one of the birds; then I lifted the hat from the other. One of the females then rushed, apparently with notes of joy and congratulation, to one of the males, who gave her a spiteful tweak and blow. Then the other came and he served her the same. He was evidently a little bewildered, and not certain what had happened or who was responsible for it. Did he think the two females were in some way to blame? But he was soon reconciled to one of [12] them again, as was the other male with the other, yet the two couples did not separate till the males had come into collision once more. Presently, however, they drifted apart, and each pair was soon holding an animated conversation punctuated by those pretty wing gestures, about the two bird-boxes.

These scenes of love and rivalry had lasted nearly all the forenoon, and matters between the birds apparently remained as they were before—the members of each pair quite satisfied with each other. One pair occupied one of the bird-boxes in the vineyard and reared two broods there during the season, but the other pair drifted away and took up their abode somewhere else.

[13]

THE BLUEBIRD

> A wistful note from out the sky,
> "Pure, pure, pure," in plaintive tone,

As if the wand'rer were alone,
And hardly knew to sing or cry.

But now a flash of eager wing,
Flitting, twinkling by the wall,
And pleadings sweet and am'rous call, —
Ah, now I know his heart doth sing!

O bluebird, welcome back again,
Thy azure coat and ruddy vest
Are hues that April loveth best, —
Warm skies above the furrowed plain.

The farm boy hears thy tender voice,
And visions come of crystal days,
With sugar-camps in maple ways,
And scenes that make his heart rejoice.

The lucid smoke drifts on the breeze,
The steaming pans are mantling white,
And thy blue wing's a joyous sight,
Among the brown and leafless trees.

[14] Now loosened currents glance and run,
And buckets shine on sturdy boles,
The forest folk peep from their holes,
And work is play from sun to sun.

The downy beats his sounding limb,
The nuthatch pipes his nasal call,
And Robin perched on tree-top tall
Heavenward lifts his evening hymn.

Now go and bring thy homesick bride,
Persuade her here is just the place
To build a home and found a race
In Downy's cell, my lodge beside.

[15]

THE ROBIN

Not long after the bluebird comes the robin. In large numbers they scour the fields and groves. You hear their piping in the meadow, in the pasture, on the hillside. Walk in the woods, and the dry leaves rustle with the whir of their wings, the air is vocal with their cheery call. In excess of joy and vivacity, they run, leap, scream, chase each other through the air, diving and sweeping among the trees with perilous rapidity.

In that free, fascinating, half-work-and-half-play pursuit,—sugar-making,—a pursuit which still lingers in many parts of New York, as in New England,—the robin is one's constant companion. When the day is sunny and the ground bare, you meet him at all points and hear him at all hours. At sunset, on the tops of the tall maples, with look heavenward, and in a spirit of utter abandonment, he carols his simple strain. And sitting thus amid the stark, silent trees, above the wet, cold earth, with the chill of winter still in the air, there is no fitter or sweeter songster in the whole round year. It is in keeping with the scene and the occasion. How round and genuine the notes are, and how eagerly our ears [16] drink them in! The first utterance, and the spell of winter is thoroughly broken, and the remembrance of it afar off.

One of the most graceful of warriors is the robin. I know few prettier sights than two males challenging and curveting about each other upon the grass in early spring. Their attentions to each other are so courteous and restrained. In alternate curves and graceful sallies, they pursue and circumvent each other. First one hops a few feet, then the other, each one standing erect in true military style while his fellow passes him and describes the segment of an ellipse about him, both uttering the while a fine complacent warble in a high but suppressed key. Are they lovers or enemies? the beholder wonders, until they make a spring and are beak to beak in the twinkling of an eye, and perhaps mount a few feet into the air, but rarely actually deliver blows upon each other. Every thrust is parried, every movement met. They follow each other with dignified composure about the fields or lawn, into trees and upon the ground, with plumage slightly spread, breasts glowing, their lisping, shrill war-song just audible. It forms on the whole the most civil and high-bred tilt to be witnessed during the season.

In the latter half of April, we pass through what I call the "robin racket,"—trains of three [17] or four birds rushing pell-mell over the lawn and fetching up in a tree or bush, or occasionally upon the ground, all piping and screaming at the top of their voices, but whether in mirth or anger it is hard to tell. The nucleus of the train is a female. One cannot see that the males in pursuit of her are rivals; it seems rather as if they had united to hustle her out of the place. But somehow the matches are no doubt made and sealed during these mad rushes. Maybe the female shouts out to her suitors, "Who touches me first wins," and away she scurries like an arrow. The males shout out, "Agreed!" and away they go in pursuit, each trying to outdo the other. The game is a brief one. Before one can get the clew to it, the party has dispersed.

The first year of my cabin life a pair of robins attempted to build a nest upon the round timber that forms the plate under my porch roof. But it was a poor place to build in. It took nearly a week's time and caused the birds a great waste of labor to find this out. The coarse material they brought for the foundation would not bed well upon the rounded surface of the timber, and every vagrant breeze that came along swept it off. My porch was kept littered with twigs and weed-stalks for days, till finally the birds abandoned [18] the undertaking. The next season a wiser or more experienced pair made the attempt again, and succeeded. They placed the nest against the rafter where it joins the plate; they used mud from the start to level up with and to hold the first twigs and straws, and had soon completed a firm, shapely structure. When the young were about ready to fly, it was interesting to note that there was apparently an older and a younger, as in most families. One bird was more advanced than any of the others. Had the parent birds intentionally stimulated it with extra quantities of food, so as to be able to launch their offspring into the world one at a time? At any rate, one of the birds was ready to leave the nest a day and a half before any of the others. I happened to be looking at it when the first impulse to get outside the nest seemed to seize it. Its parents were encouraging it with calls and assurances from some rocks a few yards away. It answered their calls in vigorous, strident tones. Then it climbed over the edge of the nest upon the plate, took a few steps forward, then a few more, till it was a yard from the nest and near the end of the timber, and could look off into free space. Its parents apparently shouted, "Come on!" But its courage was not quite equal to the leap; it looked around, and, seeing how far it was from home, [19] scampered back to the nest, and climbed into it like a frightened child. It had made its first journey into the world, but the home tie had brought it quickly back. A few hours afterward it journeyed to the end of the plate again, and then turned and rushed back. The third time its heart was braver, its wings stronger, and, leaping into the air with a shout, it flew easily to some rocks a dozen or more yards away. Each of the young in succession, at intervals of nearly a day, left the nest in this manner. There would be the first journey of a few feet along the plate, the first sudden panic at being so far from home, the rush back, a second and perhaps a third attempt, and then the irrevocable leap into the air, and a clamorous flight to a near-by bush or rock. Young birds never go back when they have once taken flight. The first free flap of the wings severs forever the ties that bind them to home.

I recently observed a robin boring for grubs in a country dooryard. It is a common enough sight to witness one seize an angleworm and drag it from its burrow in the turf, but I am not sure that I ever before saw one drill for grubs and bring the big white morsel to the surface. The robin I am speaking of had a nest of young in a maple near by, and she worked the neighborhood [20] very industriously for food. She would run along over the short grass after the manner of robins, stopping every few feet, her form stiff and erect. Now and then she would suddenly bend her head toward the ground and bring eye or ear for a moment to bear intently upon it. Then she would spring to boring the turf vigorously with her bill, changing her attitude at each stroke, alert and watchful, throwing up the grass roots and little jets of soil, stabbing deeper and deeper, growing every moment more and more excited, till finally a fat grub was seized and brought forth. Time after time, during several days, I saw her mine for grubs in this way and drag them forth. How did she know where to drill? The insect was in every case an inch below the surface. Did she hear it gnawing the roots of the grasses, or did she see a movement in the turf beneath which the grub was at work? I know not. I only know that she struck her game unerringly each time. Only twice did I see her make a few thrusts and then desist, as if she had been for the moment deceived.

[21]

THE FLICKER

Another April comer, who arrives shortly after Robin Redbreast, with whom he associates both at this season and in the autumn, is the golden-winged woodpecker, *alias* "high-hole," *alias* "flicker," *alias* "yarup," *alias* "yellow-hammer." He is an old favorite of my boyhood, and his note to me means very much. He announces his arrival by a long, loud call, repeated from the dry branch of some tree, or a stake in the fence, — a thoroughly melodious April sound. I think how Solomon finished that beautiful description of spring, "and the voice of the turtle is heard in our land," and see that a description of spring in this farming country, to be equally characteristic, should culminate in like manner, — "and the call of the high-hole comes up from the wood." It is a loud, strong, sonorous call, and does not seem to imply an answer, but rather to subserve some purpose of love or music. It is "Yarup's" proclamation of peace and good-will to all.

I recall an ancient maple standing sentry to a large sugar-bush, that, year after year, afforded protection to a brood of yellow-hammers in its decayed heart. A week or two before the nesting seemed actually to have begun, three or four of these birds might be seen, on almost any bright morning, gamboling and courting amid its decayed branches. Sometimes you would hear only a gentle persuasive cooing, or a quiet confidential chattering; then that long, loud call, taken up by first one, then another, as they sat about upon the naked limbs; anon, a sort of wild, rollicking laughter, intermingled with various cries, yelps, and squeals, as if some incident had excited their mirth and ridicule. Whether this social hilarity and boisterousness is in celebration of the pairing or mating ceremony, or whether it is only a sort of annual "house-warming" common among high-holes on resuming their summer quarters, is a question upon which I reserve my judgment.

FLICKER

Unlike most of his kinsmen, the golden-wing prefers the fields and the borders of the forest to the deeper seclusion of the woods, and hence, contrary to the habit of his tribe, obtains most of his subsistence from the ground, probing it for ants and crickets. He is not quite satisfied with being a woodpecker. He courts the society of the robin and the finches, abandons the trees for the meadow, and feeds eagerly upon berries and grain. What may be the final upshot of this course of living is a question worthy the attention of Darwin. Will his taking to the ground and his pedestrian feats result in lengthening his legs, his feeding upon berries and grains subdue his tints and soften his voice, and his associating with Robin put a song into his heart?

In the cavity of an apple-tree, much nearer the house than they usually build, a pair of high-holes took up their abode. A knot-hole which led to the decayed interior was enlarged, the live wood being cut away as clean as a squirrel would have done it. The inside preparations I could not witness, but day after day, as I passed near, I heard the bird hammering away, evidently beating down obstructions and shaping and enlarging the cavity. The chips were not brought out, but were used rather to floor the interior. The woodpeckers are not nest-builders, but rather nest-carvers.

The time seemed very short before the voices of the young were heard in the heart of the old tree,—at first feebly, but waxing stronger day by day until they could be heard many rods distant. When I put my hand upon the trunk of the tree, they would set up an eager, expectant chattering; but if I climbed up it toward the opening, they soon detected the unusual sound and would hush quickly, only now and then uttering a warning note. Long before they were fully fledged they clambered up to the orifice to receive their food. As but one could stand in the opening at a time, there was a good deal of elbowing and struggling for this position. It was a very desirable one aside from the advantages it had when food was served; it looked out upon the great, shining world, into which the young birds seemed never tired of gazing. The fresh air must have been a consideration also, for the interior of a high-hole's dwelling is not sweet. When the parent birds came with food, the young one in the opening did not get it all, but after he had received a portion, either on his own motion or on a hint from the old one, he would give place to the one behind him. Still, one bird evidently outstripped his fellows, and in the race of life was two or three days in advance of them. His voice was loudest and his head oftenest at the window. But I noticed that, when he had kept the position too long, the others evidently made it uncomfortable in his rear, and, after "fidgeting" about awhile, he would be compelled to "back down." But retaliation was then easy, and I fear his mates spent few easy moments at that lookout. They would close their eyes and slide back into the cavity as if the world had suddenly lost all its charms for them. [25]

This bird was, of course, the first to leave the nest. For two days before that event he kept his position in the opening most of the time and sent forth his strong voice incessantly. The old ones abstained from feeding him almost entirely, no doubt to encourage his exit. As I stood looking at him one afternoon and noting his progress, he suddenly reached a resolution,—seconded, I have no doubt, from the rear,—and launched forth upon his untried wings. They served him well, and carried him about fifty yards up-hill the first heat. The second day after, the next in size and spirit left in the same manner; then another, till only one remained. The parent birds ceased their visits to him, and for one day he called and called till our ears were tired of the sound. His was the faintest heart of all. Then he had none to encourage him from behind. He left the nest and clung to the outer bole of the tree, and yelped and piped for an hour longer; then he committed himself to his wings and went his way like the rest.

The matchmaking of the high-holes, which often comes under my observation, is in marked contrast to that of the robins and the bluebirds. There does not appear to be any anger or any blows. The male or two males will alight on a limb in front of the female, and go through [26] with a series of bowings and scrapings that are truly comical. He spreads his tail, he puffs out his breast, he throws back his head and then bends his body to the right and to the left, uttering all the while a curious musical hiccough. The female confronts him unmoved, but whether her attitude is critical or defensive, I cannot tell. Presently she flies away, followed by her suitor or suitors, and the little comedy is enacted on another stump or tree. Among all the woodpeckers the drum plays an important part in the matchmaking. The male takes up his stand on a dry, resonant limb, or on the ridgeboard of a building, and beats the loudest call he is capable of. A favorite drum of the high-holes about me is a hollow wooden tube, a section of a pump, which stands as a birdbox upon my summer-house. It is a good instrument; its tone is sharp and clear. A high-hole alights upon it, and sends forth a rattle that can be heard a long way off. Then he lifts up his head and utters that long April call, *Wick, wick, wick, wick.* Then he drums again. If the female does not find him, it is not because he does not make noise enough. But his sounds are all welcome to the ear. They are simple and primitive, and voice well a certain sentiment of the April days. As I write these lines I hear through the half-open [27] door his call come up from a distant field. Then I hear the steady hammering of one that has been for three days trying to penetrate the weather boarding of the big icehouse by the river, and to reach the sawdust filling for a nesting-place.

[28]

THE PHŒBE

Another April bird whose memory I fondly cherish is the phœbe-bird, the pioneer of the flycatchers. In the inland farming districts, I used to notice him, on some bright morning about Easter Day, proclaiming his arrival, with much variety of motion and attitude, from the peak of the barn or hay-shed. As yet, you may have heard only the plaintive, homesick note of the bluebird, or the faint trill of the song sparrow; and the phœbe's clear, vivacious assurance of his veritable bodily presence among us again is welcomed by all ears. At agreeable intervals in his lay he describes a circle or an ellipse in the air, ostensibly prospecting for insects, but really, I suspect, as an artistic flourish, thrown in to make up in some way for the deficiency of his musical performance. If plainness of dress indicates powers of song, as it usually does, the phœbe ought to be unrivaled in musical ability, for surely that ashen-gray suit is the superlative of plainness; and that form, likewise, would hardly pass for a "perfect figure" of a bird. The seasonableness of his coming, however, and his civil, neighborly ways, shall make up for all deficiencies in song and plumage. [29]

The phœbe-bird is a wise architect and perhaps enjoys as great an immunity from danger, both in its person and its nest, as any other bird. Its modest ashen-gray suit is the color of the rocks where it builds, and the moss of which it makes such free use gives to its nest the look of a natural growth or accretion. But when it comes into the barn or under the shed to build, as it so frequently does, the moss is rather out of place. Doubtless in time the bird will take the hint, and when she builds in such places will leave the moss out. I noted but two nests the summer I am speaking of: one in a barn failed of issue, on account of the rats, I suspect, though the little owl may have been the depredator; the other, in the woods, sent forth three young. This latter nest was most charmingly and ingeniously placed. I discovered it while in quest of pond-lilies, in a long, deep, level stretch of water in the woods. A large tree had blown over at the edge of the water, and its dense mass of upturned roots, with the black, peaty soil filling the interstices, was like the fragment of a wall several feet high, rising from the edge of the languid current. In a niche in this earthy wall, and visible and accessible only from the water, a phœbe had built her nest and reared her brood. I paddled my boat up and came alongside prepared to take the family [30] aboard. The young, nearly ready to fly, were quite undisturbed by my presence, having probably been assured that no danger need be apprehended from that side. It was not a likely place for minks, or they would not have been so secure.

[31]

THE COMING OF PHŒBE

When buckets shine 'gainst maple trees
And drop by drop the sap doth flow,
When days are warm, but still nights freeze,
And deep in woods lie drifts of snow,
When cattle low and fret in stall,
Then morning brings the phœbe's call,
"Phœbe,
Phœbe, phœbe," a cheery note,
While cackling hens make such a rout.

When snowbanks run, and hills are bare,
And early bees hum round the hive,
When woodchucks creep from out their lair
Right glad to find themselves alive,
When sheep go nibbling through the fields,
Then Phœbe oft her name reveals,
"Phœbe,
Phœbe, phœbe," a plaintive cry,
While jack-snipes call in morning sky.

When wild ducks quack in creek and pond
And bluebirds perch on mullein-stalks,
When spring has burst her icy bond
And in brown fields the sleek crow walks,
[32] When chipmunks court in roadside walls,
Then Phœbe from the ridgeboard calls,
"Phœbe,
Phœbe, phœbe," and lifts her cap,

While smoking Dick doth boil the sap.

[33]

THE COWBIRD

The cow blackbird is a noticeable songster in April, though it takes a back seat a little later. It utters a peculiarly liquid April sound. Indeed, one would think its crop was full of water, its notes so bubble up and regurgitate, and are delivered with such an apparent stomachic contraction. This bird is the only feathered polygamist we have. The females are greatly in excess of the males, and the latter are usually attended by three or four of the former. As soon as the other birds begin to build, they are on the *qui vive*, prowling about like gypsies, not to steal the young of others, but to steal their eggs into other birds' nests, and so shirk the labor and responsibility of hatching and rearing their own young.

The cowbird's tactics are probably to watch the movements of the parent bird. She may often be seen searching anxiously through the trees or bushes for a suitable nest, yet she may still oftener be seen perched upon some good point of observation watching the birds as they come and go about her. There is no doubt that, in many cases, the cowbird makes room for her own illegitimate [34] egg in the nest by removing one of the bird's own. I found a sparrow's nest with two sparrow's eggs and one cowbird's egg, and another egg lying a foot or so below it on the ground. I replaced the ejected egg, and the next day found it again removed, and another cowbird's egg in its place. I put it back the second time, when it was again ejected, or destroyed, for I failed to find it anywhere. Very alert and sensitive birds, like the warblers, often bury the strange egg beneath a second nest built on top of the old. A lady living in the suburbs of an Eastern city heard cries of distress one morning from a pair of house wrens that had a nest in a honeysuckle on her front porch. On looking out of the window, she beheld this little comedy,—comedy from her point of view, but no doubt grim tragedy from the point of view of the wrens: a cowbird with a wren's egg in its beak running rapidly along the walk, with the outraged wrens forming a procession behind it, screaming, scolding, and gesticulating as only these voluble little birds can. The cowbird had probably been surprised in the act of violating the nest, and the wrens were giving her a piece of their minds.

Every cowbird is reared at the expense of two or more song-birds. For every one of these dusky [35] little pedestrians there amid the grazing cattle there are two or more sparrows, or vireos, or warblers, the less. It is a big price to pay,—two larks for a bunting,—two sovereigns for a shilling; but Nature does not hesitate occasionally to contradict herself in just this way. The young of the cowbird is disproportionately large and aggressive, one might say hoggish. When disturbed, it will clasp the nest and scream and snap its beak threateningly. One was hatched out in a song sparrow's nest which was under my observation, and would soon have overridden and overborne the young sparrow which came out of the shell a few hours later, had I not interfered from time to time and lent the young sparrow a helping hand. Every day I would visit the nest and take the sparrow out from under the potbellied interloper, and place it on top, so that presently it was able to hold its own against its enemy. Both birds became fledged and left the nest about the same time. Whether the race was an even one after that, I know not.

[36]

THE CHIPPING SPARROW

When the true flycatcher catches a fly, it is quick business. There is no strife, no pursuit,—one fell swoop, and the matter is ended. Now note that yonder little sparrow is less skilled. It is the chippy, and he finds his subsistence properly in various seeds and the larvæ of insects, though he occasionally has higher aspirations, and seeks to emulate the pewee, commencing and ending his career as a flycatcher by an awkward chase after a beetle or "miller." He is hunting around in the grass now, I suspect, with the desire to indulge this favorite whim. There!—the opportunity is afforded him. Away goes a little cream-colored meadow-moth in the most tortuous course he is capable of, and away goes Chippy in pursuit. The contest is quite comical, though I dare say it is serious enough to the moth. The chase continues for a few yards, when there is a sudden rushing to cover in the grass,—then a taking to wing again, when the search has become too close, and the moth has recovered his wind. Chippy chirps angrily, and is determined not to be beaten. Keeping, with the slightest effort, upon the heels of the fugitive, he is ever [37] on the point of halting to snap him up, but never quite does it; and so, between disappointment and expectation, is soon disgusted, and returns to pursue his more legitimate means of subsistence.

Last summer I made this record in my notebook: "A nest of young robins in the maple in front of the house being fed by a chipping sparrow. The little sparrow is very attentive; seems decidedly fond of her adopted babies. The old robins resent her services, and hustle her out of the tree whenever they find her near the nest. (It was this hurried departure of Chippy from the tree that first attracted my attention.) She watches her chances, and comes with food in their absence. The young birds are about ready to fly, and when the chippy feeds them her head fairly disappears in their capacious mouths. She jerks it back as if she were afraid of being swallowed. Then she lingers near them on the edge of the nest, and seems to admire them. When she sees the old robin coming, she spreads her wings in an attitude of defense, and then flies away. I wonder if she has had the experience of rearing a cow-bunting?" (A day later.) "The robins are out of the nest, and the little sparrow continues to feed them. She approaches them rather timidly [38] and hesitatingly, as if she feared they might swallow her, then thrusts her titbit quickly into the distended mouth and jerks back."

Whether the chippy had lost her own brood, whether she was an unmated bird, or whether the case was simply the overflowing of the maternal instinct, it would be interesting to know.

[39]

THE CHEWINK

CHEWINK
Upper, male; lower, female

The chewink is a shy bird, but not stealthy. It is very inquisitive, and sets up a great scratching among the leaves, apparently to attract your attention. The male is perhaps the most conspicuously marked of all the ground-birds except the bobolink, being black above, bay on the sides, and white beneath. The bay is in compliment to the leaves he is forever scratching among,—they have rustled against his breast and sides so long that these parts have taken their color; but whence come the white and the black? The bird seems to be aware that his color betrays him, for there are few birds in the woods so careful about keeping themselves screened from view. When in song, its favorite perch is the top of some high bush near to cover. On being disturbed at such times, it pitches down into the brush and is instantly lost to view.

This is the bird that Thomas Jefferson wrote to Wilson about, greatly exciting the latter's curiosity. Wilson was just then upon the threshold of his career as an ornithologist, and had made a drawing of the Canada jay which he sent to the President. It was a new bird, and in reply [40] Jefferson called his attention to a "curious bird" which was everywhere to be heard, but scarcely ever to be seen. He had for twenty years interested the young sportsmen of his neighborhood to shoot one for him, but without success. "It is in all the forests, from spring to fall," he says in his letter, "and never but on the tops of the tallest trees, from which it perpetually serenades us with some of the sweetest notes, and as clear as those of the nightingale. I have followed it for miles, without ever but once getting a good view of it. It is of the size and make of the mockingbird, lightly thrush-colored on the back, and a grayish-white on the breast and belly. Mr. Randolph, my son-in-law, was in possession of one which had been shot by a neighbor," etc. Randolph pronounced it a flycatcher, which was a good way wide of the mark. Jefferson must have seen only the female, after all his tramp, from his description of the color; but he was doubtless following his own great thoughts more than the bird, else he would have had an earlier view. The bird was not a new one, but was well known then as the ground-robin. The President put Wilson on the wrong scent by his erroneous description, and it was a long time before the latter got at the truth of the case. But Jefferson's letter is a good sample of [41] those which specialists often receive from intelligent persons who have seen or heard something in their line very curious or entirely new, and who set the man of science agog by a description of the supposed novelty,—a description that generally fits the facts of the case about as well as your coat fits the chair-back. Strange and curious things in the air, and in the water, and in the earth beneath, are seen every day except by those who are looking for them, namely, the naturalists. When Wilson or Audubon gets his eye on the unknown bird, the illusion vanishes, and your phenomenon turns out to be one of the commonplaces of the fields or woods.

[42]

THE BROWN THRASHER

Our long-tailed thrush, or thrasher, delights in a high branch of some solitary tree, whence it will pour out its rich and intricate warble for an hour together. This bird is the great American chipper. There is no other bird that I know of that can chip with such emphasis and military decision as this yellow-eyed songster. It is like the click of a giant gunlock. Why is the thrasher so stealthy? It always seems to be going about on tip-toe. I never knew it to steal anything, and yet it skulks and hides like a fugitive from justice. One never sees it flying aloft in the air and traversing the world openly, like most birds, but it darts along fences and through bushes as if pursued by a guilty conscience. Only when the musical fit is upon it does it come up into full view, and invite the world to hear and behold.

Years pass without my finding a brown thrasher's nest; it is not a nest you are likely to stumble upon in your walk; it is hidden as a miser hides his gold, and watched as jealously. The male pours out his rich and triumphant song from the tallest tree he can find, and fairly challenges you to [43] come and look for his treasures in his vicinity. But you will not find them if you go. The nest is somewhere on the outer circle of his song; he is never so imprudent as to take up his stand very near it. The artists who draw those cozy little pictures of a brooding mother bird, with the male perched but a yard away in full song, do not copy from nature. The thrasher's nest I found was thirty or forty rods from the point where the male was wont to indulge in his brilliant recitative. It was in an open field under a low ground-juniper. My dog disturbed the sitting bird as I was passing near. The nest could be seen only by lifting up and parting away the branches. All the arts of concealment had been carefully studied. It was the last place you would think of looking in, and, if you did look, nothing was visible but the dense green circle of the low-spreading juniper. When you approached, the bird would keep her place till you had begun to stir the branches, when she would start out, and, just skimming the ground, make a bright brown line to the near fence and bushes. I confidently expected that this nest would escape molestation, but it did not. Its discovery by myself and dog probably opened the door for ill luck, as one day, not long afterward, when I peeped in upon it, it was empty. The proud song of the male had ceased from his accustomed [44] tree, and the pair were seen no more in that vicinity.

After a pair of nesting birds have been broken up once or twice during the season, they become almost desperate, and will make great efforts to outwit their enemies. A pair of brown thrashers built their nest in a pasture-field under a low, scrubby apple-tree which the cattle had browsed down till it spread a thick, wide mass of thorny twigs only a few inches above the ground. Some blackberry briers had also grown there, so that the screen was perfect. My dog first started the bird, as I was passing near. By stooping low and peering intently, I could make out the nest and eggs. Two or three times a week, as I passed by, I would pause to see how the nest was prospering. The mother bird would keep her place, her yellow eyes never blinking. One morning, as I looked into her tent, I found the nest empty. Some night-prowler, probably a skunk or a fox, or maybe a black snake or a red squirrel by day, had plundered it. It would seem as if it was too well screened; it was in such a spot as any depredator would be apt to explore. "Surely," he would say, "this is a likely place for a nest." The birds then moved over the hill a hundred rods or more, much nearer the house, and in some rather open bushes tried again. But again they [45] came to grief. Then, after some delay, the mother bird made a bold stroke. She seemed to reason with herself thus: "Since I have fared so disastrously in seeking seclusion for my nest, I will now adopt the opposite tactics, and come out fairly in the open. What hides me hides my enemies: let us try greater publicity." So she came out and built her nest by a few small shoots that grew beside the path that divides the two vineyards, and where we passed to and fro many times daily. I discovered her by chance early in the morning as I proceeded to my work. She started up at my feet and flitted quickly along above the ploughed ground, almost as red as the soil. I admired her audacity. Surely no prowler by night or day would suspect a nest in this open and exposed place. There was no cover by which they could approach, and no concealment anywhere. The nest was a hasty affair, as if the birds' patience at nest-building had been about exhausted. Presently an egg appeared, and then the next day another, and on the fourth day a third. No doubt the bird would have succeeded this time had not man interfered. In cultivating the vineyards the horse and cultivator had to pass over this very spot. Upon this the bird had not calculated. I determined to assist her. I called my man, and told him there was one spot [46] in that vineyard, no bigger than his hand, where the horse's foot must not be allowed to fall, nor tooth of cultivator to touch. Then I showed him the nest, and charged him to avoid it. Probably if I had kept the secret to myself, and let the bird run her own risk, the nest would have escaped. But the result was that the man, in elaborately trying to avoid the nest, overdid the matter; the horse plunged, and set his foot squarely upon it. Such a little spot, the chances were few that the horse's foot would fall exactly there; and yet it did, and the birds' hopes were again dashed. The pair then disappeared from my vicinity, and I saw them no more.

[47]

THE HOUSE WREN

A few years ago I put up a little bird-house in the back end of my garden for the accommodation of the wrens, and every season a pair have taken up their abode there. One spring a pair of bluebirds looked into the tenement and lingered about several days, leading me to hope that they would conclude to occupy it. But they finally went away, and later in the season the wrens appeared, and, after a little coquetting, were regularly installed in their old quarters, and were as happy as only wrens can be.

One of our younger poets, Myron Benton, saw a little bird

"Ruffled with whirlwind of his ecstasies,"

which must have been the wren, as I know of no other bird that so throbs and palpitates with music as this little vagabond. And the pair I speak of seemed exceptionally happy, and the male had a small tornado of song in his crop that kept him "ruffled" every moment in the day. But before their honeymoon was over the bluebirds returned. I knew something was wrong before I was up in the morning. Instead of that voluble and gushing song outside the window, I heard the [48] wrens scolding and crying at a fearful rate, and on going out saw the bluebirds in possession of the box. The poor wrens were in despair; they wrung their hands and tore their hair, after the wren fashion, but chiefly did they rattle out their disgust and wrath at the intruders. I have no doubt that, if it could have been interpreted, it would have been proven the rankest and most voluble billingsgate ever uttered. For the wren is saucy, and he has a tongue in his head that can outwag any other tongue known to me.

The bluebirds said nothing, but the male kept an eye on Mr. Wren, and, when he came too near, gave chase, driving him to cover under the fence, or under a rubbish-heap or other object, where the wren would scold and rattle away, while his pursuer sat on the fence or the pea-brush waiting for him to reappear.

Days passed, and the usurpers prospered and the outcasts were wretched; but the latter lingered about, watching and abusing their enemies, and hoping, no doubt, that things would take a turn, as they presently did. The outraged wrens were fully avenged. The mother bluebird had laid her full complement of eggs and was beginning to set, when one day, as her mate was perched above her on the barn, along came a boy with one of those wicked elastic slings and cut him down [49] with a pebble. There he lay like a bit of sky fallen upon the grass. The widowed bird seemed to understand what had happened, and without much ado disappeared next day in quest of another mate.

In the mean time the wrens were beside themselves with delight; they fairly screamed with joy. If the male was before "ruffled with whirlwind of his ecstasies," he was now in danger of being rent asunder. He inflated his throat and caroled as wren never caroled before. And the female, too, how she cackled and darted about! How busy they both were! Rushing into the nest, they hustled those eggs out in less than a minute, wren time. They carried in new material, and by the third day were fairly installed again in their old quarters; but on the third day, so rapidly are these little dramas played, the female bluebird reappeared with another mate. Ah! how the wren stock went down then! What dismay and despair filled again those little breasts! It was pitiful. They did not scold as before, but after a day or two withdrew from the garden, dumb with grief, and gave up the struggle.

The chatter of a second brood of nearly fledged wrens is heard now (August 20) in an oriole's nest suspended from the branch of an apple-tree [50] near where I write. Earlier in the season the parent birds made long and determined attempts to establish themselves in a cavity that had been occupied by a pair of bluebirds. The original proprietor of the place was the downy woodpecker. He had excavated it the autumn before, and had passed the winter there, often to my certain knowledge lying abed till nine o'clock in the morning. In the spring he went elsewhere, probably with a female, to begin the season in new quarters. The bluebirds early took possession, and in June their first brood had flown. The wrens had been hanging around, evidently with an eye on the place (such little comedies may be witnessed anywhere), and now very naturally thought it was their turn. A day or two after the young bluebirds had flown, I noticed some fine, dry grass clinging to the entrance to the cavity; a circumstance which I understood a few moments later, when the wren rushed by me into the cover of a small Norway spruce, hotly pursued by the male bluebird. It was a brown streak and a blue streak pretty close together. The wrens had gone to housecleaning, and the bluebird had returned to find his bed and bedding being pitched out of doors, and had thereupon given the wrens to understand in the most emphatic manner that he had no intention [51] of vacating the premises so early in the season. Day after day, for more than two weeks, the male bluebird had to clear his premises of these intruders. It occupied much of his time and not a little of mine, as I sat with a book in a summer-house near by, laughing at his pretty fury and spiteful onset. On two occasions the wren rushed under the chair in which I sat, and a streak of blue lightning almost flashed in my very face. One day, just as I had passed the tree in which the cavity was located, I heard the wren scream desperately; turning, I saw the little vagabond fall into the grass with the wrathful bluebird fairly upon him; the latter had returned just in time to catch him, and was evidently bent on punishing him well. But in the squabble in the grass the wren escaped and took refuge in the friendly evergreen. The bluebird paused for a moment with outstretched wings looking for the fugitive, then flew away. A score of times during the month of June did I see the wren taxing every energy to get away from the bluebird. He would dart into the stone wall, under the floor of the summer-house, into the weeds,—anywhere to hide his diminished head. The bluebird, with his bright coat, looked like an officer in uniform in pursuit of some wicked, rusty little street gamin. Generally the favorite house of refuge [52] of the wrens was the little spruce, into which their pursuer made no attempt to follow them. The female would sit concealed amid the branches, chattering in a scolding, fretful way, while the male with his eye upon his tormentor would perch on the topmost shoot and sing. Why he sang at such times, whether in triumph and derision, or to keep his courage up and reassure his mate, I could not make out. When his song was suddenly cut short, and I glanced to see him dart down into the spruce, my eye usually caught a twinkle of blue wings hovering near. The wrens finally gave up the fight, and their enemies reared their second brood in peace.

[53]

THE SONG SPARROW

The first song sparrow's nest I observed in the spring of 1881 was in a field under a fragment of a board, the board being raised from the ground a couple of inches by two poles. It had its full complement of eggs, and probably sent forth a brood of young birds, though as to this I cannot speak positively, as I neglected to observe it further. It was well sheltered and concealed, and was not easily come at by any of its natural enemies, save snakes and weasels. But concealment often avails little. In May, a song sparrow, which had evidently met with disaster earlier in the season, built its nest in a thick mass of woodbine against the side of my house, about fifteen feet from the ground. Perhaps it took the hint from its cousin the English sparrow. The nest was admirably placed, protected from the storms by the overhanging eaves and from all eyes by the thick screen of leaves. Only by patiently watching the suspicious bird, as she lingered near with food in her beak, did I discover its whereabouts. That brood is safe, I thought, beyond doubt. But it was not: the nest was pillaged one night, either by an owl, or else by a rat that had climbed into [54] the vine, seeking an entrance to the house. The mother bird, after reflecting upon her ill luck about a week, seemed to resolve to try a different system of tactics, and to throw all appearances of concealment aside. She built a nest a few yards from the house, beside the drive, upon a smooth piece of greensward. There was not a weed or a shrub or anything whatever to conceal it or mark its site. The structure was completed, and incubation had begun, before I discovered what was going on. "Well, well," I said, looking down upon the bird almost at my feet, "this is going to the other extreme indeed; now the cats will have you." The desperate little bird sat there day after day, looking like a brown leaf pressed down in the short green grass. As the weather grew hot, her position became very trying. It was no longer a question of keeping the eggs warm, but of keeping them from roasting. The sun had no mercy on her, and she fairly panted in the middle of the day. In such an emergency the male robin has been known to perch above the sitting female and shade her with his outstretched wings. But in this case there was no perch for the male bird, had he been disposed to make a sunshade of himself. I thought to lend a hand in this direction myself, and so stuck a leafy twig beside the nest. This was probably an unwise [55] interference: it guided disaster to the spot; the nest was broken up, and the mother bird was probably caught, as I never saw her afterward.

One day a tragedy was enacted a few yards from where I was sitting with a book: two song sparrows were trying to defend their nest against a black snake. The curious, interrogating note of a chicken who had suddenly come upon the scene in his walk first caused me to look up from my reading. There were the sparrows, with wings raised in a way peculiarly expressive of horror and dismay, rushing about a low clump of grass and bushes. Then, looking more closely, I saw the glistening form of the black snake, and the quick movement of his head as he tried to seize the birds. The sparrows darted about and through the grass and weeds, trying to beat the snake off. Their tails and wings were spread, and, panting with the heat and the desperate struggle, they presented a most singular spectacle. They uttered no cry, not a sound escaped them; they were plainly speechless with horror and dismay. Not once did they drop their wings, and the peculiar expression of those uplifted palms, as it were, I shall never forget. It occurred to me that perhaps here was a case of attempted bird-charming on the part of the snake, so I looked on from behind the fence. The birds charged the [56] snake and harassed him from every side, but were evidently under no spell save that of courage in defending their nest. Every moment or two I could see the head and neck of the serpent make a sweep at the birds, when the one struck at would fall back, and the other would renew the assault from the rear. There appeared to be little danger that the snake could strike and hold one of the birds, though I trembled for them, they were so bold and approached so near to the snake's head. Time and again he sprang at them, but without success. How the poor things panted, and held up their wings appealingly! Then the snake glided off to the near fence, barely escaping the stone which I hurled at him. I found the nest rifled and deranged; whether it had contained eggs or young, I know not. The male sparrow had cheered me many a day with his song, and I blamed myself for not having rushed at once to the rescue, when the arch enemy was upon him. There is probably little truth in the popular notion that snakes charm birds. The black snake is the most subtle, alert, and devilish of our snakes, and I have never seen him have any but young, helpless birds in his mouth.

If one has always built one's nest upon the ground, and if one comes of a race of ground-builders, [57] it is a risky experiment to build in a tree. The conditions are vastly different. One of my near neighbors, a little song sparrow, learned this lesson the past season. She grew ambitious; she departed from the traditions of her race, and placed her nest in a tree. Such a pretty spot she chose, too,—the pendent cradle formed by the interlaced sprays of two parallel branches of a Norway spruce. These branches shoot out almost horizontally; indeed, the lower ones become quite so in spring, and the side shoots with which they are clothed droop down, forming the slopes of miniature ridges; where the slopes of two branches join, a little valley is formed, which often looks more stable than it really is. My sparrow selected one of these little valleys about six feet from the ground, and quite near the walls of the house. "Here," she thought, "I will build my nest, and pass the heat of June in a miniature Norway. This tree is the fir-clad mountain, and this little vale on its side I select for my own." She carried up a great quantity of coarse grass and straws for the foundation, just as she would have done upon the ground. On the top of this mass there gradually came into shape the delicate structure of her nest, compacting and refining till its delicate carpet of hairs and threads was reached. So sly as the little bird was about it, too,—every moment [58] on her guard lest you discover her secret! Five eggs were laid, and incubation was far advanced, when the storms and winds came. The cradle indeed did rock. The boughs did not break, but they swayed and separated as you would part your two interlocked hands. The ground of the little valley fairly gave way, the nest tilted over till its contents fell into the chasm. It was like an earthquake that destroys a hamlet.

No born tree-builder would have placed its nest in such a situation. Birds that build at the end of the branch, like the oriole, tie the nest fast; others, like the robin, build against the main trunk; still others build securely in the fork. The sparrow, in her ignorance, rested her house upon the spray of two branches, and when the tempest came, the branches parted company and the nest was engulfed.

A little bob-tailed song sparrow built her nest in a pile of dry brush very near the kitchen door of a farmhouse on the skirts of the northern Catskills, where I was passing the summer. It was late in July, and she had doubtless reared one brood in the earlier season. Her toilet was decidedly the worse for wear. I noted her day after day, very busy about the fence and quince bushes between the house and milk house, with her beak full of coarse straw and hay. To a casual observer, [59] she seemed flitting about aimlessly, carrying straws from place to place just to amuse herself. When I came to watch her closely to learn the place of her nest, she seemed to suspect my intention, and made many little feints and movements calculated to put me off my track. But I would not be misled, and presently had her secret. The male did not assist her at all, but sang much of the time in an apple-tree or upon the fence, on the other side of the house.

The song sparrow nearly always builds upon the ground, but my little neighbor laid the foundations of her domicile a foot or more above the soil. And what a mass of straws and twigs she did collect together! How coarse and careless and aimless at first,—a mere lot of rubbish dropped upon the tangle of dry limbs; but presently how it began to refine and come into shape in the centre! till there was the most exquisite hair-lined cup set about by a chaos of coarse straws and branches. What a process of evolution! The completed nest was foreshadowed by the first stiff straw; but how far off is yet that dainty casket with its complement of speckled eggs! The nest was so placed that it had for canopy a large, broad, drooping leaf of yellow dock. This formed a perfect shield against both sun and rain, while it served to conceal it from [60] any curious eyes from above,—from the cat, for instance, prowling along the top of the wall. Before the eggs had hatched, the docken leaf wilted and dried and fell down upon the nest. But the mother bird managed to insinuate herself beneath it, and went on with her brooding all the same.

Then I arranged an artificial cover of leaves and branches, which shielded her charge till they had flown away. A mere trifle was this little bob-tailed bird with her arts and her secrets, and the male with his song, and yet the pair gave a touch of something to those days and to that place which I would not willingly have missed.

[61]

THE CHIMNEY SWIFT

One day a swarm of honey-bees went into my chimney, and I mounted the stack to see into which flue they had gone. As I craned my neck above the sooty vent, with the bees humming about my ears, the first thing my eye rested upon in the black interior was a pair of long white pearls upon a little shelf of twigs, the nest of the chimney swallow, or swift,—honey, soot, and birds' eggs closely associated. The bees, though in an unused flue, soon found the gas of anthracite that hovered about the top of the chimney too much for them, and they left. But the swifts are not repelled by smoke. They seem to have entirely abandoned their former nesting-places in hollow trees and stumps, and to frequent only chimneys. A tireless bird, never perching, all day upon the wing, and probably capable of flying one thousand miles in twenty-four hours, they do not even stop to gather materials for their nests, but snap off the small dry twigs from the tree-tops as they fly by. Confine one of these swifts to a room and it does not perch, but after flying till it becomes bewildered and exhausted, it clings to the side of the wall till it [62] dies. Once, on returning to my room after several days' absence, I found one in which life seemed nearly extinct; its feet grasped my finger as I removed it from the wall, but its eyes closed, and it seemed about on the point of joining its companion, which lay dead upon the floor. Tossing it into the air, however, seemed to awaken its wonderful powers of flight, and away it went straight toward the clouds. On the wing the chimney swift looks like an athlete stripped for the race. There is the least appearance of quill and plumage of any of our birds, and, with all its speed and marvelous evolutions, the effect of its flight is stiff and wiry. There appears to be but one joint in the wing, and that next the body. This peculiar inflexible motion of the wings, as if they were little sickles of sheet iron, seems to be owing to the length and development of the primary quills and the smallness of the secondary. The wing appears to hinge only at the wrist. The barn swallow lines its rude masonry with feathers, but the swift begins life on bare twigs, glued together by a glue of home manufacture as adhesive as Spaulding's.

The big chimney of my cabin "Slabsides" of course attracted the chimney swifts, and as it was not used in summer, two pairs built their nests in it, and we had the muffled thunder of their [63] wings at all hours of the day and night. One night, when one of the broods was nearly fledged, the nest that held them fell down into the fireplace. Such a din of screeching and chattering as they instantly set up! Neither my dog nor I could sleep. They yelled in chorus, stopping at the end of every half-minute as if upon signal. Now they were all screeching at the top of their voices, then a sudden, dead silence ensued. Then the din began again, to terminate at the instant as before. If they had been long practicing together, they could not have succeeded better. I never before heard the cry of birds so accurately timed. After a while I got up and put them back up the chimney, and stopped up the throat of the flue with newspapers. The next day one of the parent birds, in bringing food to them, came down the chimney with such force that it passed through the papers and brought up in the fireplace. On capturing it I saw that its throat was distended with food as a chipmunk's cheek with corn, or a boy's pocket with chestnuts. I opened its mandibles, when it ejected a wad of insects as large as a bean. Most of them were much macerated, but there were two house-flies yet alive and but little the worse for their close confinement. They stretched themselves and walked about upon my hand, enjoying a breath [64] of fresh air once more. It was nearly two hours before the swift again ventured into the chimney with food.

These birds do not perch, nor alight upon buildings or the ground. They are apparently upon the wing all day. They outride the storms. I have in my mind a cheering picture of three of them I saw facing a heavy thunder-shower one afternoon. The wind was blowing a gale, the clouds were rolling in black, portentous billows out of the west, the peals of thunder were shaking the heavens, and the big drops were just beginning to come down, when, on looking up, I saw three swifts high in air, working their way slowly, straight into the teeth of the storm. They were not hurried or disturbed; they held themselves firmly and steadily; indeed, they were fairly at anchor in the air till the rage of the elements should have subsided. I do not know that any other of our land birds outride the storms in this way.

In the choice of nesting-material the swift shows no change of habit. She still snips off the small dry twigs from the tree-tops and glues them together, and to the side of the chimney, with her own glue. The soot is a new obstacle in her way, that she does not yet seem to have learned to overcome, as the rains often loosen it and [65] cause her nest to fall to the bottom. She has a pretty way of trying to frighten you off when your head suddenly darkens the opening above her. At such times she leaves the nest and clings to the side of the chimney near it. Then, slowly raising her wings, she suddenly springs out from the wall and back again, making as loud a drumming with them in the passage as she is capable of. If this does not frighten you away, she repeats it three or four times. If your face still hovers above her, she remains quiet and watches you.

What a creature of the air this bird is, never touching the ground, so far as I know, and never tasting earthly food! The swallow does perch now and then and descend to the ground for nesting-material, but not so the swift. The twigs for her nest she gathers on the wing, sweeping along like children on a "merry-go-round" who try to seize a ring, or to do some other feat, as they pass a given point. If the swift misses the twig, or it fails to yield to her the first time, she tries again and again, each time making a wider circuit, as if to tame and train her steed a little and bring him up more squarely to the mark next time.

Though the swift is a stiff flyer and apparently without joints in her wings, yet the air of frolic and of superabundance of wing-power is [66] more marked with her than with any other of our birds. Her feeding and twig-gathering seem like asides in a life of endless play. Several times both in spring and fall I have seen swifts gather in immense numbers toward nightfall, to take refuge in large unused chimney-stacks. On such occasions they seem to be coming together for some aerial festival or grand celebration; and, as if bent upon a final effort to work off a part of their superabundant wing-power before settling down for the night, they circle and circle high above the chimney-top, a great cloud of them, drifting this way and that, all in high spirits and chippering as they fly. Their numbers constantly increase as other members of the clan come dashing in from all points of the compass. Swifts seem to materialize out of empty air on all sides of the chippering, whirling ring, as an hour or more this assembling of the clan and this flight festival go on. The birds must gather in from whole counties, or from half a State. They have been on the wing all day, and yet now they seem as tireless as the wind, and as if unable to curb their powers.

One fall they gathered in this way and took refuge for the night in a large chimney-stack in a city near me, and kept this course up for more than a month and a half. Several times I went [67] to town to witness the spectacle, and a spectacle it was: ten thousand swifts, I should think, filling the air above a whole square like a whirling swarm of huge black bees, but saluting the ear with a multitudinous chippering, instead of a humming. People gathered upon the sidewalks to see them. It was a rare circus performance, free to all. After a great many feints and playful approaches, the whirling ring of birds would suddenly grow denser above the chimney; then a stream of them, as if drawn down by some power of suction, would pour into the opening. For only a few seconds would this downward rush continue; then, as if the spirit of frolic had again got the upper hand of them, the ring would rise, and the chippering and circling go on. In a minute or two the same manœuvre would be repeated, the chimney, as it were, taking its swallows at intervals to prevent choking. It usually took a half-hour or more for the birds all to disappear down its capacious throat. There was always an air of timidity and irresolution about their approach to the chimney, just as there always is about their approach to the dead tree-top from which they procure their twigs for nest-building. Often did I see birds hesitate above the opening and then pass on, apparently as though they had not struck it at just the right angle. [68] On one occasion a solitary bird was left flying, and it took three or four trials either to make up its mind or to catch the trick of the descent. On dark or threatening or stormy days the birds would begin to assemble by mid-afternoon, and by four or five o'clock were all in their lodgings.

[69]

THE OVEN-BIRD

Every loiterer about the woods knows this pretty, speckled-breasted, olive-backed little bird, which walks along over the dry leaves a few yards from him, moving its head as it walks, like a miniature domestic fowl. Most birds are very stiff-necked, like the robin, and as they run or hop upon the ground, carry the head as if it were riveted to the body. Not so the oven-bird, or the other birds that walk, as the cow-bunting, or the quail, or the crow. They move the head forward with the movement of the feet. The sharp, reiterated, almost screeching song of the oven-bird, as it perches on a limb a few feet from the ground, like the words "preacher, preacher, preacher," or "teacher, teacher, teacher," uttered louder and louder, and repeated six or seven times, is also familiar to most ears; but its wild, ringing, rapturous burst of song in the air high above the tree-tops is not so well known. From a very prosy, tiresome, unmelodious singer, it is suddenly transformed for a brief moment into a lyric poet of great power. It is a great surprise. The bird undergoes a complete transformation. Ordinarily it is a very quiet, demure sort of bird. It walks [70] about over the leaves, moving its head like a little hen; then perches on a limb a few feet from the ground and sends forth its shrill, rather prosy, unmusical chant. Surely it is an ordinary, commonplace bird. But wait till the inspiration of its flight-song is upon it. What a change! Up it goes through the branches of the trees, leaping from limb to limb, faster and faster, till it shoots from the tree-tops fifty or more feet into the air above them, and bursts into an ecstasy of song, rapid, ringing, lyrical; no more like its habitual performance than a match is like a rocket; brief but thrilling; emphatic but musical. Having reached its climax of flight and song, the bird closes its wings and drops nearly perpendicularly downward like the skylark. If its song were more prolonged, it would rival the song of that famous bird. The bird does this many times a day during early June, but oftenest at twilight.

About the first of June there is a nest in the woods, upon the ground, with four creamy-white eggs in it, spotted with brown or lilac, chiefly about the larger ends, that always gives the walker who is so lucky as to find it a thrill of pleasure. It is like a ground sparrow's nest with a roof or canopy to it. The little brown or olive backed bird starts away from your feet and runs swiftly and almost silently over the dry leaves, [71] and then turns her speckled breast to see if you are following. She walks very prettily, by far the prettiest pedestrian in the woods. But if she thinks you have discovered her secret, she feigns lameness and disability of both leg and wing, to decoy you into the pursuit of her. This is the oven-bird. The last nest of this bird I found was while in quest of the pink cypripedium. We suddenly spied a couple of the flowers a few steps from the path along which we were walking, and had stooped to admire them, when out sprang the bird from beside them, doubtless thinking she was the subject of observation instead of the rose-purple flowers that swung but a foot or two above her. But we never should have seen her had she kept her place. She had found a rent in the matted carpet of dry leaves and pine needles that covered the ground, and into this had insinuated her nest, the leaves and needles forming a canopy above it, sloping to the south and west, the source of the more frequent summer rains.

[72]

THE CATBIRD

It requires an effort for me to speak of the singing catbird as he; all the ways and tones of the bird seem so distinctly feminine. But it is, of course, only the male that sings. At times I hardly know whether I am more pleased or annoyed with him. Perhaps he is a little too common, and his part in the general chorus a little too conspicuous. If you are listening for the note of another bird, he is sure to be prompted to the most loud and protracted singing, drowning all other sounds; if you sit quietly down to observe a favorite or study a new-comer, his curiosity knows no bounds, and you are scanned and ridiculed from every point of observation. Yet I would not miss him; I would only subordinate him a little, make him less conspicuous.

He is the parodist of the woods, and there is ever a mischievous, bantering, half-ironical undertone in his lay, as if he were conscious of mimicking and disconcerting some envied songster. Ambitious of song, practicing and rehearsing in private, he yet seems the least sincere and genuine of the sylvan minstrels, as if he had taken up music only to be in the fashion, or not to be outdone [73] by the robins and thrushes. In other words, he seems to sing from some outward motive, and not from inward joyousness. He is a good versifier, but not a great poet. Vigorous, rapid, copious, not without fine touches, but destitute of any high, serene melody, his performance, like that of Thoreau's squirrel, always implies a spectator.

There is a certain air and polish about his strain, however, like that in the vivacious conversation of a well-bred lady of the world, that commands respect. His parental instinct, also, is very strong, and that simple structure of dead twigs and dry grass is the centre of much anxious solicitude. Not long since, while strolling through the woods, my attention was attracted to a small densely-grown swamp, hedged in with eglantine, brambles, and the everlasting smilax, from which proceeded loud cries of distress and alarm, indicating that some terrible calamity was threatening my sombre-colored minstrel. On effecting an entrance, which, however, was not accomplished till I had doffed coat and hat, so as to diminish the surface exposed to the thorns and brambles, and, looking around me from a square yard of terra firma, I found myself the spectator of a loathsome yet fascinating scene. Three or four yards from me was the nest, beneath which, [74] in long festoons, rested a huge black snake; a bird two-thirds grown was slowly disappearing between his expanded jaws. As he seemed unconscious of my presence, I quietly observed the proceedings. By slow degrees he compassed the bird about with his elastic mouth; his head flattened, his neck writhed and swelled, and two or three undulatory movements of his glistening body finished the work. Then he cautiously raised himself up, his tongue flaming from his mouth the while, curved over the nest, and, with wavy, subtle motions, explored the interior. I can conceive of nothing more overpoweringly terrible to an unsuspecting family of birds than the sudden appearance above their domicile of the head and neck of this arch-enemy. It is enough to petrify the blood in their veins. Not finding the object of his search, he came streaming down from the nest to a lower limb, and commenced extending his researches in other directions, sliding stealthily through the branches, bent on capturing one of the parent birds. That a legless, wingless creature should move with such ease and rapidity where only birds and squirrels are considered at home, lifting himself up, letting himself down, running out on the yielding boughs, and traversing with marvelous celerity the whole length and breadth of the thicket, was truly surprising. One [75] thinks of the great myth of the Tempter and the "cause of all our woe," and wonders if the Arch Enemy is not now playing off some of his pranks before him. Whether we call it snake or devil matters little. I could but admire his terrible beauty, however; his black, shining folds, his easy, gliding movement, head erect, eyes glistening, tongue playing like subtle flame, and the invisible means of his almost winged locomotion.

The parent birds, in the mean while, kept up the most agonizing cry, at times fluttering furiously about their pursuer, and actually laying hold of his tail with their beaks and claws. On being thus attacked, the snake would suddenly double upon himself and follow his own body back, thus executing a strategic movement that at first seemed almost to paralyze his victim and place her within his grasp. Not quite, however. Before his jaws could close upon the coveted prize the bird would tear herself away, and, apparently faint and sobbing, retire to a higher branch. His reputed powers of fascination availed him little, though it is possible that a frailer and less combative bird might have been held by the fatal spell. Presently, as he came gliding down the slender body of a leaning alder, his attention was attracted by a slight movement of my arm; eyeing me an instant, with that crouching, utterly [76] motionless gaze which I believe only snakes and devils can assume, he turned quickly — a feat which necessitated something like crawling over his own body — and glided off through the branches, evidently recognizing in me a representative of the ancient parties he once so cunningly ruined. A few moments later, as he lay carelessly disposed in the top of a rank alder, trying to look as much like a crooked branch as his supple, shining form would admit, the old vengeance overtook him. I exercised my prerogative, and a well-directed missile, in the shape of a stone, brought him looping and writhing to the ground. After I had completed his downfall and quiet had been partly restored, a half-fledged member of the bereaved household came out from his hiding-place, and, jumping upon a decayed branch, chirped vigorously, no doubt in celebration of the victory.

[77]

THE BOBOLINK

The bobolink has a secure place in literature, having been laureated by no less a poet than Bryant, and invested with a lasting human charm in the sunny page of Irving, and is the only one of our songsters, I believe, that the mockingbird cannot parody or imitate. He affords the most marked example of exuberant pride, and a glad, rollicking, holiday spirit, that can be seen among our birds. Every note expresses complacency and glee. He is a beau of the first pattern, and, unlike any other bird of my acquaintance, pushes his gallantry to the point of wheeling gayly into the train of every female that comes along, even after the season of courtship is over and the matches are all settled; and when she leads him on too wild a chase, he turns lightly about and breaks out with a song that is precisely analogous to a burst of gay and self-satisfied laughter, as much as to say, "*Ha! ha! ha! I must have my fun, Miss Silverthimble, thimble, thimble, if I break every heart in the meadow, see, see, see!*"

At the approach of the breeding-season the bobolink undergoes a complete change; his form changes, his color changes, his flight changes. [78] From mottled brown or brindle he becomes black and white, earning, in some localities, the shocking name of "skunk bird"; his small, compact form becomes broad and conspicuous, and his ordinary flight is laid aside for a mincing, affected gait, in which he seems to use only the very tips of his wings. It is very noticeable what a contrast he presents to his mate at this season, not only in color but in manners, she being as shy and retiring as he is forward and hilarious. Indeed, she seems disagreeably serious and indisposed to any fun or jollity, scurrying away at his approach, and apparently annoyed at every endearing word and look. It is surprising that all this parade of plumage and tinkling of cymbals should be gone through with and persisted in to please a creature so coldly indifferent as she really seems to be.

I know of no other song-bird that expresses so much self-consciousness and vanity, and comes so near being an ornithological coxcomb. The redbird, the yellowbird, the indigo-bird, the oriole, the cardinal grosbeak, and others, all birds of brilliant plumage and musical ability, seem quite unconscious of self, and neither by tone nor act challenge the admiration of the beholder.

If I were a bird, in building my nest I should follow the example of the bobolink, placing it in [79] the midst of a broad meadow, where there was no spear of grass, or flower, or growth unlike another to mark its site. I judge that the bobolink escapes the dangers to which nesting birds are liable as few or no other birds do. Unless the mowers come along at an earlier date than she has anticipated, that is, before July 1, or a skunk goes nosing through the grass, which is unusual, she is as safe as bird well can be in the great open of nature. She selects the most monotonous and uniform place she can find amid the daisies or the timothy and clover, and places her simple structure upon the ground in the midst of it. There is no concealment, except as the great conceals the little, as the desert conceals the pebble, as the myriad conceals the unit. You may find the nest once, if your course chances to lead you across it, and your eye is quick enough to note the silent brown bird as she darts swiftly away; but step three paces in the wrong direction, and your search will probably be fruitless. My friend and I found a nest by accident one day, and then lost it again one minute afterward. I moved away a few yards to be sure of the mother bird, charging my friend not to stir from his tracks. When I returned, he had moved two paces, he said (he had really moved four), and we spent a half-hour stooping [80] over the daisies and the buttercups, looking for the lost clew. We grew desperate, and fairly felt the ground over with our hands, but without avail. I marked the spot with a bush, and came the next day, and, with the bush as a centre, moved about it in slowly increasing circles, covering, I thought, nearly every inch of ground with my feet, and laying hold of it with all the visual power I could command, till my patience was exhausted, and I gave up, baffled. I began to doubt the ability of the parent birds themselves to find it, and so secreted myself and watched. After much delay, the male bird appeared with food in his beak, and, satisfying himself that the coast was clear, dropped into the grass which I had trodden down in my search. Fastening my eye upon a particular meadow-lily, I walked straight to the spot, bent down, and gazed long and intently into the grass. Finally my eye separated the nest and its young from its surroundings. My foot had barely missed them in my search, but by how much they had escaped my eye I could not tell. Probably not by distance at all, but simply by unrecognition. They were virtually invisible. The dark gray and yellowish-brown dry grass and stubble of the meadow-bottom were exactly copied in the color of the half-fledged young. More than that, they hugged the nest so closely and formed such a compact mass, that though there were five of them, they preserved the unit of expression,—no single head or form was defined; they were one, and that one was without shape or color, and not separable, except by closest scrutiny, from the one of the meadow-bottom. That nest prospered, as bobolinks' nests doubtless generally do; for, notwithstanding the enormous slaughter of the birds by Southern sportsmen during their fall migrations, the bobolink appears to hold its own, and its music does not diminish in our Northern meadows.

THE BOBOLINK

>Daisies, clover, buttercup,
>Redtop, trefoil, meadowsweet,
>Ecstatic pinions, soaring up,
>Then gliding down to grassy seat.
>
>Sunshine, laughter, mad desires,
>May day, June day, lucid skies,
>All reckless moods that love inspires—
>The gladdest bird that sings and flies.
>
>Meadows, orchards, bending sprays,
>Rushes, lilies, billowy wheat,
>Song and frolic fill his days,
>A feathered rondeau all complete.
>
>Pink bloom, gold bloom, fleabane white,
>Dewdrop, raindrop, cooling shade,
>Bubbling throat and hovering flight,
>And jocund heart as e'er was made.

[83]

THE WOOD THRUSH

The wood thrush is the handsomest species of the thrush family. In grace and elegance of manner he has no equal. Such a gentle, high-bred air, and such inimitable ease and composure in his flight and movement! He is a poet in very word and deed. His carriage is music to the eye. His performance of the commonest act, as catching a beetle, or picking a worm from the mud, pleases like a stroke of wit or eloquence. Was he a prince in the olden time, and do the regal grace and mien still adhere to him in his transformation? What a finely proportioned form! How plain, yet rich, his color,—the bright russet of his back, the clear white of his breast, with the distinct heart-shaped spots! It may be objected to Robin that he is noisy and demonstrative; he hurries away or rises to a branch with an angry note, and flirts his wings in ill-bred suspicion. The thrasher, or red thrush, sneaks and skulks like a culprit, hiding in the densest alders; the catbird is a coquette and a flirt, as well as a sort of female Paul Pry; and the chewink shows his inhospitality by espying your movements like a detective. The wood thrush [84] has none of these underbred traits. He regards me unsuspiciously, or avoids me with a noble reserve—or, if I am quiet and incurious, graciously hops toward me, as if to pay his respects, or to make my acquaintance. I have passed under his nest within a few feet of his mate and brood, when he sat near by on a branch eying me sharply, but without opening his beak; but the moment I raised my hand toward his defenseless household his anger and indignation were beautiful to behold.

What a noble pride he has! Late one October, after his mates and companions had long since gone South, I noticed one for several successive days in the dense part of this next-door wood, flitting noiselessly about, very grave and silent, as if doing penance for some violation of the code of honor. By many gentle, indirect approaches, I perceived that part of his tail-feathers were undeveloped. The sylvan prince could not think of returning to court in this plight, and so, amid the falling leaves and cold rains of autumn, was patiently biding his time.

WOOD THRUSH

It is a curious habit the wood thrush has of starting its nest with a fragment of newspaper or other paper. Except in remote woods, I think it nearly always puts a piece of paper in the foundation of its nest. Last spring I chanced to be sitting [85] near a tree in which a wood thrush had concluded to build. She came with a piece of paper nearly as large as my hand, placed it upon the branch, stood upon it a moment, and then flew down to the ground. A little puff of wind caused the paper to leave the branch a moment afterward. The thrush watched it eddy slowly down to the ground, when she seized it and carried it back. She placed it in position as before, stood upon it again for a moment, and then flew away. Again the paper left the branch, and sailed away slowly to the ground. The bird seized it again, jerking it about rather spitefully, I thought; she turned it round two or three times, then labored back to the branch with it, upon which she shifted it about as if to hit upon some position in which it would lie more securely. This time she sat down upon it for a moment, and then went away, doubtless with the thought in her head that she would bring something to hold it down. The perverse paper followed her in a few seconds. She seized it again, and hustled it about more than before. As she rose with it toward the nest, it in some way impeded her flight, and she was compelled to return to the ground with it. But she kept her temper remarkably well. She turned the paper over and took it up in her beak several times before she was satisfied with her hold, and [86] then carried it back to the branch, where, however, it would not stay. I saw her make six trials of it, when I was called away. I think she finally abandoned the restless fragment, probably a scrap that held some "breezy" piece of writing, for later in the season I examined the nest and found no paper in it.

How completely the life of a bird revolves about its nest, its home! In the case of the wood thrush, its life and joy seem to mount higher and higher as the nest prospers. The male becomes a fountain of melody; his happiness waxes day by day; he makes little triumphal tours about the neighborhood, and pours out his pride and gladness in the ears of all. How sweet, how well-bred, is his demonstration! But let any accident befall that precious nest, and what a sudden silence falls upon him! Last summer a pair of wood thrushes built their nest within a few rods of my house, and when the enterprise was fairly launched and the mother bird was sitting upon her four blue eggs, the male was in the height of his song. How he poured forth his rich melody, never in the immediate vicinity of the nest, but always within easy hearing distance! Every morning, as promptly as the morning came, between five and six, he would sing for half an hour from the top of a locust-tree that shaded [87] my roof. I came to expect him as much as I expected my breakfast, and I was not disappointed till one morning I seemed to miss something. What was it? Oh, the thrush had not sung this morning. Something is the matter; and, recollecting that yesterday I had seen a red squirrel in the trees not far from the nest, I at once inferred that the nest had been harried. Going to the spot, I found my fears were well grounded; every egg was gone. The joy of the thrush was laid low. No more songs from the tree-top, and no more songs from any point, till nearly a week had elapsed, when I heard him again under the hill, where the pair had started a new nest, cautiously tuning up, and apparently with his recent bitter experience still weighing upon him.

There is no nest-builder that suffers more from crows and squirrels and other enemies than the wood thrush. It builds as openly and unsuspiciously as if it thought all the world as honest as itself. Its favorite place is the fork of a sapling, eight or ten feet from the ground, where it falls an easy prey to every nest-robber that comes prowling through the woods and groves. It is not a bird that skulks and hides, like the catbird, the brown thrasher, the chat, or the chewink, and its nest is not concealed with the same art as theirs. Our thrushes are all frank, open-mannered [88] birds; but the veery and the hermit build on the ground, where they may at least escape the crows, owls, and jays, and stand a good chance of being overlooked by the red squirrel and weasel also; while the robin seeks the protection of dwellings and outbuildings. For years I have not known the nest of a wood thrush to succeed. During the season referred to I observed but two, both apparently a second attempt, as the season was well advanced, and both failures. In one case, the nest was placed in a branch that an apple-tree, standing near a dwelling, held out over the highway. The structure was barely ten feet above the middle of the road, and would just escape a passing load of hay. It was made conspicuous by the use of a large fragment of newspaper in its foundation,—an unsafe material to build upon in most cases. Whatever else the press may guard, this particular newspaper did not guard this nest from harm. It saw the egg and probably the chick, but not the fledgeling. A murderous deed was committed above the public highway, but whether in the open day or under cover of darkness I have no means of knowing. The frisky red squirrel was doubtless the culprit. The other nest was in a maple sapling, within a few yards of the little rustic summer-house already referred to. The first attempt [89] of the season, I suspect, had failed in a more secluded place under the hill; so the pair had come up nearer the house for protection. The male sang in the trees near by for several days before I chanced to see the nest. The very morning, I think, it was finished, I saw a red squirrel exploring a tree but a few yards away; he probably knew what the singing meant as well as I did. I did not see the inside of the nest, for it was almost instantly deserted, the female having probably laid a single egg, which the squirrel had devoured.

One evening, while seated upon my porch, I had convincing proof that musical or song contests do take place among the birds. Two wood thrushes who had nests near by sat on the top of a dead tree and pitted themselves against each other in song for over half an hour, contending like champions in a game, and certainly affording the rarest treat in wood-thrush melody I had ever had. They sang and sang with unwearied spirit and persistence, now and then changing position or facing in another direction, but keeping within a few feet of each other. The rivalry became so obvious and was so interesting that I finally made it a point not to take my eyes from the singers. The twilight deepened till their forms began to grow dim; then one of the birds [90] could stand the strain no longer, the limit of fair competition had been reached, and seeming to say, "I will silence you, anyhow," it made a spiteful dive at its rival, and in hot pursuit the two disappeared in the bushes beneath the tree.

[91]

THE BALTIMORE ORIOLE

The nest of nests, the ideal nest, is unquestionably that of the Baltimore oriole. It is the only perfectly pensile nest we have. The nest of the orchard oriole is indeed mainly so, but this bird generally builds lower and shallower, more after the manner of the vireos.

The Baltimore oriole loves to attach its nest to the swaying branches of the tallest elms, making no attempt at concealment, but satisfied if the position be high and the branch pendent. This nest would seem to cost more time and skill than any other bird structure. A peculiar flax-like material seems to be always sought after and always found. The nest when completed assumes the form of a large, suspended gourd. The walls are thin but firm, and proof against the most driving rain. The mouth is hemmed or over-handed with strings or horsehair, and the sides are usually sewed through and through with the same.

BALTIMORE ORIOLE
Upper, male; lower, female

Not particular as to the matter of secrecy, the bird is not particular as to material, so that it be of the nature of strings or threads. A lady friend once told me that, while she was working [92] by an open window, one of these birds approached while her back was turned, and, seizing a skein of some kind of thread or yarn, made off with it to its half-finished nest. But the perverse yarn caught fast in the branches, and, in the bird's efforts to extricate it, got hopelessly tangled. She tugged away at it all day, but was finally obliged to content herself with a few detached portions. The fluttering strings were an eyesore to her ever after, and, passing and repassing, she would give them a spiteful jerk, as much as to say, "There is that confounded yarn that gave me so much trouble."

One day in Kentucky I saw an oriole weave into her nest unusual material. As we sat upon the lawn in front of the cottage, we had noticed the bird just beginning her structure, suspending it from a long, low branch of the Kentucky coffee tree that grew but a few feet away. I suggested to my host that if he would take some brilliant yarn and scatter it about upon the shrubbery, the fence, and the walks, the bird would probably avail herself of it, and weave a novel nest. I had heard of its being done, but had never tried it myself. The suggestion was at once acted upon, and in a few moments a handful of zephyr yarn, crimson, orange, green, yellow, and blue, was distributed about the grounds. [93] As we sat at dinner a few moments later, I saw the eager bird flying up toward her nest with one of these brilliant yarns streaming behind her. They had caught her eye at once, and she fell to work upon them with a will; not a bit daunted by their brilliant color, she soon had a crimson spot there amid the green leaves. She afforded us rare amusement all the afternoon and the next morning. How she seemed to congratulate herself over her rare find! How vigorously she knotted those strings to her branch and gathered the ends in and sewed them through and through the structure, jerking them spitefully like a housewife burdened with many cares! How savagely she would fly at her neighbor, an oriole that had a nest just over the fence a few yards away, when she invaded her territory! The male looked on approvingly, but did not offer to lend a hand. There is something in the manner of the female on such occasions, something so decisive and emphatic, that one entirely approves of the course of the male in not meddling or offering any suggestions. It is the wife's enterprise, and she evidently knows her own mind so well that the husband keeps aloof, or plays the part of an approving spectator.

The woolen yarn was ill-suited to the Kentucky climate. This fact the bird seemed to appreciate, [94] for she used it only in the upper part of her nest, in attaching it to the branch and in binding and compacting the rim, making the sides and bottom of hemp, leaving it thin and airy, much more so than are the same nests with us. No other bird would, perhaps, have used such brilliant material; their instincts of concealment would have revolted, but the oriole aims more to make its nest inaccessible than to hide it. Its position and depth insure its safety.

[95]

THE WHIP-POOR-WILL

One day in May, walking in the woods, I came upon the nest of a whip-poor-will, or rather its eggs, for it builds no nest,—two elliptical whitish spotted eggs lying upon the dry leaves. My foot was within a yard of the mother bird before she flew. I wondered what a sharp eye would detect curious or characteristic in the ways of the bird, so I came to the place many times and had a look. It was always a task to separate the bird from her surroundings, though I stood within a few feet of her, and knew exactly where to look. One had to bear on with his eye, as it were, and refuse to be baffled. The sticks and leaves, and bits of black or dark-brown bark, were all exactly copied in the bird's plumage. And then she did sit so close, and simulate so well a shapeless, decaying piece of wood or bark! Twice I brought a companion, and, guiding his eye to the spot, noted how difficult it was for him to make out there, in full view upon the dry leaves, any semblance to a bird. When the bird returned after being disturbed, she would alight within a few inches of her eggs, and then, after a moment's pause, hobble awkwardly upon them. [96]

WHIP-POOR-WILL

After the young had appeared, all the wit of the bird came into play. I was on hand the next day, I think. The mother bird sprang up when I was within a pace of her, and in doing so fanned the leaves with her wings till they sprang up, too; as the leaves started the young started, and as they were of the same color, to tell which was the leaf and which the bird was a trying task to any eye. I came the next day, when the same tactics were repeated. Once a leaf fell upon one of the young birds and nearly hid it. The young are covered with a reddish down, like a young partridge, and soon follow their mother about. When disturbed, they gave but one leap, then settled down, perfectly motionless and stupid, with eyes closed. The parent bird, on these occasions, made frantic efforts to decoy me away from her young. She would fly a few paces and fall upon her breast, and a spasm, like that of death, would run through her tremulous outstretched wings and prostrate body. She kept a sharp eye out the mean while to see if the ruse took, and, if it did not, she was quickly cured, and, moving about to some other point, tried to draw my attention as before. When followed she always alighted upon the ground, dropping down in a sudden, peculiar way. The second or third day both old and young had disappeared. [97]

The whip-poor-will walks as awkwardly as a swallow, which is as awkward as a man in a bag, and yet she manages to lead her young about the woods. The latter, I think, move by leaps and sudden spurts, their protective coloring shielding them most effectively.

As the shadows deepen and the stars begin to come out, the whip-poor-will suddenly strikes up. What a rude intrusion upon the serenity and harmony of the hour! A cry without music, insistent, reiterated, loud, penetrating, and yet the ear welcomes it; the night and the solitude are so vast that they can stand it; and when, an hour later, as the night enters into full possession, the bird comes and serenades me under my window or upon my doorstep, my heart warms toward it. Its cry is a love-call, and there is something of the ardor and persistence of love in it, and when the female responds, and comes and hovers near, there is an interchange of subdued, caressing tones between the two birds that it is a delight to hear. During my first summer in my cabin one bird used to strike up every night from a high ledge of rocks in front of my door. At just such a moment in the twilight he would begin, the first to break the stillness. Then the others would follow, till the solitude was vocal with [98] their calls. They are rarely heard later than ten o'clock. Then at daybreak they take up the tale again, whipping poor Will till one pities him. One April morning between three and four o'clock, hearing one strike up near my window, I began counting its calls. My neighbor had told me he had heard one call over two hundred times without a break, which seemed to me a big story. But I have a much bigger one to tell. This bird actually laid upon the back of poor Will one thousand and eighty-eight blows, with only a barely perceptible pause here and there, as if to catch its breath. Then it stopped about half a minute and began again, uttering this time three hundred and ninety calls, when it paused, flew a little farther away, took up the tale once more, and continued till I fell asleep.

By day the whip-poor-will apparently sits motionless upon the ground. A few times in my walks through the woods I have started one up from almost under my feet. On such occasions the bird's movements suggest those of a bat; its wings make no noise, and it wavers about in an uncertain manner, and quickly drops to the ground again. One June day we flushed an old one with her two young, but there was no indecision or hesitation in the manner of the mother bird this time. The young were more than half [99] fledged, and they scampered away a few yards and suddenly squatted upon the ground, where their assimilative coloring rendered them almost invisible. Then the anxious parent put forth all her arts to absorb our attention and lure us away from her offspring. She flitted before us from side to side, with spread wings and tail, now falling upon the ground, where she would remain a moment as if quite disabled, then perching upon an old stump or low branch with drooping, quivering wings, and imploring us by every gesture to take her and spare her young. My companion had his camera with him, but the bird would not remain long enough in one position for him to get her picture.

[100]

THE BLACK-THROATED BLUE WARBLER
A SEARCH FOR A RARE NEST

I had set out in hopes of finding a rare nest,—the nest of the black-throated blue-backed warbler, which, it seemed, with one or two others, was still wanting to make the history of our warblers complete. The woods were extensive, and full of deep, dark tangles, and looking for any particular nest seemed about as hopeless a task as searching for a needle in a haystack, as the old saying is. Where to begin, and how? But the principle is the same as in looking for a hen's nest,—first find your bird, then watch its movements.

The bird is in these woods, for I have seen him scores of times, but whether he builds high or low, on the ground or in the trees, is all unknown to me. That is his song now,—"twe-twea-twe-e-e-a," with a peculiar summer languor and plaintiveness, and issuing from the lower branches and growths. Presently we—for I have been joined by a companion—discover the bird, a male, insecting in the top of a newly fallen hemlock. The black, white, and blue of his uniform [101] are seen at a glance. His movements are quite slow compared with some of the warblers. If he will only betray the locality of that little domicile where his plainly clad mate is evidently sitting, it is all we will ask of him. But this he seems in no wise disposed to do. Here and there, and up and down, we follow him, often losing him, and as often refinding him by his song; but the clew to his nest, how shall we get it? Does he never go home to see how things are getting on, or to see if his presence is not needed, or to take madam a morsel of food? No doubt he keeps within earshot, and a cry of distress or alarm from the mother bird would bring him to the spot in an instant. Would that some evil fate would make her cry, then! Presently he encounters a rival. His feeding-ground infringes upon that of another, and the two birds regard each other threateningly. This is a good sign, for their nests are evidently near.

Their battle-cry is a low, peculiar chirp, not very fierce, but bantering and confident. They quickly come to blows, but it is a very fantastic battle, and, as it would seem, indulged in more to satisfy their sense of honor than to hurt each other, for neither party gets the better of the other, and they separate a few paces and sing, and squeak, and challenge each other in a very happy [102] frame of mind. The gauntlet is no sooner thrown down than it is again taken up by one or the other, and in the course of fifteen or twenty minutes they have three or four encounters, separating a little, then provoked to return again like two cocks, till finally they withdraw beyond hearing of each other,—both, no doubt, claiming the victory. But the secret of the nest is still kept. Once I think I have it. I catch a glimpse of a bird which looks like the female, and near by, in a small hemlock about eight feet from the ground, my eye detects a nest. But as I come up under it, I can see daylight through it, and that it is empty,—evidently only partly finished, not lined or padded yet. Now if the bird will only return and claim it, the point will be gained. But we wait and watch in vain. The architect has knocked off to-day, and we must come again, or continue our search.

Despairing of finding either of the nests of the two males, we pushed on through the woods to try our luck elsewhere. Before long, just as we were about to plunge down a hill into a dense, swampy part of the woods, we discovered a pair of the birds we were in quest of. They had food in their beaks, and, as we paused, showed great signs of alarm, indicating that the nest was in the immediate vicinity. This was enough. We [103] would pause here and find this nest, anyhow. To make a sure thing of it, we determined to watch the parent birds till we had wrung from them their secret. So we doggedly crouched down and watched them, and they watched us. It was diamond cut diamond. But as we felt constrained in our movements, desiring, if possible, to keep so quiet that the birds would, after a while, see in us only two harmless stumps or prostrate logs, we had much the worst of it. The mosquitoes were quite taken with our quiet, and knew us from logs and stumps in a moment. Neither were the birds deceived, not even when we tried the Indian's tactics, and plumed ourselves with green branches. Ah, the suspicious creatures, how they watched us with the food in their beaks, abstaining for one whole hour from ministering to that precious charge which otherwise would have been visited every few moments! Quite near us they would come at times, between us and the nest, eying us so sharply. Then they would move off, and apparently try to forget our presence. Was it to deceive us, or to persuade himself and his mate that there was no serious cause for alarm, that the male would now and then strike up in full song and move off to some distance through the trees? But the mother bird did not allow herself to lose sight of us at all, and both birds, after carrying [104] the food in their beaks a long time, would swallow it themselves. Then they would obtain another morsel and apparently approach very near the nest, when their caution or prudence would come to their aid, and they would swallow the food and hasten away. I thought the young birds would cry out, but not a syllable from them. Yet this was, no doubt, what kept the parent birds away from the nest. The clamor the young would have set up on the approach of the old with food would have exposed everything.

After a time I felt sure I knew within a few feet where the nest was concealed. Indeed, I thought I knew the identical bush. Then the birds approached each other again and grew very confidential about another locality some rods below. This puzzled us, and, seeing the whole afternoon might be spent in this manner and the mystery unsolved, we determined to change our tactics and institute a thorough search of the locality. This procedure soon brought things to a crisis, for, as my companion clambered over a log by a little hemlock, a few yards from where we had been sitting, with a cry of alarm out sprang the young birds from their nest in the hemlock, and, scampering and fluttering over the leaves, disappeared in different directions. Instantly the parent birds were on the scene in an agony of alarm. [105] Their distress was pitiful. They threw themselves on the ground at our very feet, and fluttered, and cried, and trailed themselves before us, to draw us away from the place, or distract our attention from the helpless young. I shall not forget the male bird, how bright he looked, how sharp the contrast as he trailed his painted plumage there on the dry leaves. Apparently he was seriously disabled. He would start up as if exerting every muscle to fly away, but no use; down he would come, with a helpless, fluttering motion, before he had gone two yards, and apparently you had only to go and pick him up. But before you could pick him up, he had recovered somewhat and flown a little farther; and thus, if you were tempted to follow him, you would soon find yourself some distance from the scene of the nest, and both old and young well out of your reach. The female bird was not less solicitous, and practiced the same arts upon us to decoy us away, but her dull plumage rendered her less noticeable. The male was clad in holiday attire, but his mate in an every-day working-garb.

The nest was built in the fork of a little hemlock, about fifteen inches from the ground, and was a thick, firm structure, composed of the finer material of the woods, with a lining of very delicate roots or rootlets. There were four young birds and one addled egg.

[106]

THE MARSH HAWK
A MARSH HAWK'S NEST, A YOUNG HAWK, AND A VISIT TO A QUAIL ON HER NEST

Most country boys, I fancy, know the marsh hawk. It is he you see flying low over the fields, beating about bushes and marshes and dipping over the fences, with his attention directed to the ground beneath him. He is a cat on wings. He keeps so low that the birds and mice do not see him till he is fairly upon them. The hen-hawk swoops down upon the meadow-mouse from his position high in air, or from the top of a dead tree; but the marsh hawk stalks him and comes suddenly upon him from over the fence, or from behind a low bush or tuft of grass. He is nearly as large as the hen-hawk, but has a much longer tail. When I was a boy I used to call him the long-tailed hawk. The male is of a bluish slate-color; the female reddish-brown, like the hen-hawk, with a white rump.

Unlike the other hawks, they nest on the ground in low, thick marshy places. For several seasons a pair have nested in a bushy marsh a few miles back of me, near the house of a farmer friend of mine, who has a keen eye for the wild [107] life about him. Two years ago he found the nest, but when I got over to see it the next week, it had been robbed, probably by some boys in the neighborhood. The past season, in April or May, by watching the mother bird, he found the nest again. It was in a marshy place, several acres in extent, in the bottom of a valley, and thickly grown with hardback, prickly ash, smilax, and other low thorny bushes. My friend took me to the brink of a low hill, and pointed out to me in the marsh below us, as nearly as he could, just where the nest was located. Then we crossed the pasture, entered upon the marsh, and made our way cautiously toward it. The wild, thorny growths, waist-high, had to be carefully dealt with. As we neared the spot, I used my eyes the best I could, but I did not see the hawk till she sprang into the air not ten yards away from us. She went screaming upward, and was soon sailing in a circle far above us. There, on a coarse matting of twigs and weeds, lay five snow-white eggs, a little more than half as large as hens' eggs. My companion said the male hawk would probably soon appear and join the female, but he did not. She kept drifting away to the east, and was soon gone from our sight.

We presently withdrew and secreted ourselves behind the stone wall, in hopes of seeing the mother [108] hawk return. She appeared in the distance, but seemed to know she was being watched, and kept away.

About ten days later we made another visit to the nest. An adventurous young Chicago lady also wanted to see a hawk's nest, and so accompanied us. This time three of the eggs were hatched, and as the mother hawk sprang up, either by accident or intentionally she threw two of the young hawks some feet from the nest. She rose up and screamed angrily. Then, turning toward us, she came like an arrow straight at the young lady, a bright plume in whose hat probably drew her fire. The damsel gathered up her skirts about her and beat a hasty retreat. Hawks were not so pretty as she thought they were. A large hawk launched at one's face from high in the air is calculated to make one a little nervous. It is such a fearful incline down which the bird comes, and she is aiming exactly toward your eye. When within about thirty feet of you, she turns upward with a rushing sound, and, mounting higher, falls toward you again. She is only firing blank cartridges, as it were; but it usually has the desired effect, and beats the enemy off.

After we had inspected the young hawks, a neighbor of my friend offered to conduct us to a [109] quail's nest. Anything in the shape of a nest is always welcome, it is such a mystery, such a centre of interest and affection, and, if upon the ground, is usually something so dainty and exquisite amid the natural wreckage and confusion. A ground nest seems so exposed, too, that it always gives a little thrill of pleasurable surprise to see the group of frail eggs resting there behind so slight a barrier. I will walk a long distance any day just to see a song sparrow's nest amid the stubble or under a tuft of grass. It is a jewel in a rosette of jewels, with a frill of weeds or turf. A quail's nest I had never seen, and to be shown one within the hunting-ground of this murderous hawk would be a double pleasure. Such a quiet, secluded, grass-grown highway as we moved along was itself a rare treat. Sequestered was the word that the little valley suggested, and peace the feeling the road evoked. The farmer, whose fields lay about us, half grown with weeds and bushes, evidently did not make stir or noise enough to disturb anything. Beside this rustic highway, bounded by old mossy stone walls, and within a stone's throw of the farmer's barn, the quail had made her nest. It was just under the edge of a prostrate thorn-bush.

"The nest is right there," said the farmer, [110] pausing within ten feet of it, and pointing to the spot with his stick.

In a moment or two we could make out the mottled brown plumage of the sitting bird. Then we approached her cautiously till we bent above her.

She never moved a feather.

Then I put my cane down in the brush behind her. We wanted to see the eggs, yet did not want rudely to disturb the sitting hen.

She would not move.

Then I put down my hand within a few inches of her; still she kept her place. Should we have to lift her off bodily?

Then the young lady put down her hand, probably the prettiest and the whitest hand the quail had ever seen. At least it started her, and off she sprang, uncovering such a crowded nest of eggs as I had never before beheld. Twenty-one of them! a ring or disk of white like a china tea-saucer. You could not help saying, How pretty! How cunning! like baby hens' eggs, as if the bird were playing at sitting, as children play at housekeeping.

If I had known how crowded her nest was, I should not have dared disturb her, for fear she would break some of them. But not an egg suffered harm by her sudden flight. And no harm [111] came to the nest afterward. Every egg hatched, I was told, and the little chicks, hardly bigger than bumblebees, were led away by the mother into the fields.

In about a week I paid another visit to the hawk's nest. The eggs were all hatched, and the mother bird was hovering near. I shall never forget the curious expression of those young hawks sitting there on the ground. The expression was not one of youth, but of extreme age. Such an ancient, infirm look as they had, — the sharp, dark, and shrunken look about the face and eyes, and their feeble, tottering motions! They sat upon their elbows and the hind part of their bodies, and their pale, withered legs and feet extended before them in the most helpless fashion. Their angular bodies were covered with a pale yellowish down, like that of a chicken; their heads had a plucked, seedy appearance; and their long, strong, naked wings hung down by their sides till they touched the ground: power and ferocity in the first rude draught, shorn of everything but its sinister ugliness. Another curious thing was the gradation of the young in size; they tapered down regularly from the first to the fifth, as if there had been, as probably there was, an interval of a day or two between the hatchings. [112]

The two older ones showed some signs of fear on our approach, and one of them threw himself upon his back, and put up his impotent legs, and glared at us with open beak. The two smaller ones regarded us not at all. Neither of the parent birds appeared during our stay.

When I visited the nest again, eight or ten days later, the birds were much grown, but of as marked a difference in size as before, and with the same look of extreme old age, — old age in men of the aquiline type, nose and chin coming together, and eyes large and sunken. They now glared upon us with a wild, savage look, and opened their beaks threateningly.

The next week, when my friend visited the nest, the larger of the hawks fought him savagely. But one of the brood, probably the last to hatch, had made but little growth. It appeared to be on the point of starvation. The mother hawk (for the male seemed to have disappeared) had perhaps found her family too large for her, and was deliberately allowing one of the number to perish; or did the larger and stronger young devour all the food before the weaker member could obtain any? Probably this was the case.

Arthur brought the feeble nestling away, and the same day my little boy got it and brought it home, wrapped in a woolen rag. It was clearly a [113] starved bantling. It cried feebly but would not lift up its head.

We first poured some warm milk down its throat, which soon revived it, so that it would swallow small bits of flesh. In a day or two we had it eating ravenously, and its growth became noticeable. Its voice had the sharp whistling character of that of its parents, and was stilled only when the bird was asleep. We made a pen for it, about a yard square, in one end of the study, covering the floor with several thicknesses of newspapers; and here, upon a bit of brown woolen blanket for a nest, the hawk waxed strong day by day. An uglier-looking pet, tested by all the rules we usually apply to such things, would have been hard to find. There he would sit upon his elbows, his helpless feet out in front of him, his great featherless wings touching the floor, and shrilly cry for more food. For a time we gave him water daily from a stylograph-pen filler, but the water he evidently did not need or relish. Fresh meat, and plenty of it, was his demand. And we soon discovered that he liked game, such as mice, squirrels, birds, much better than butcher's meat.

Then began a lively campaign on the part of my little boy against all the vermin and small game in the neighborhood, to keep the hawk supplied. He trapped and he hunted, he enlisted his [114] mates in his service, he even robbed the cats to feed the hawk. His usefulness as a boy of all work was seriously impaired. "Where is J— —?" "Gone after a squirrel for his hawk." And often the day would be half gone before his hunt was successful. The premises were very soon cleared of mice, and the vicinity of chipmunks and squirrels. Farther and farther he was compelled to hunt the surrounding farms and woods to keep up with the demands of the hawk. By the time the hawk was ready to fly, it had consumed twenty-one chipmunks, fourteen red squirrels, sixteen mice, and twelve English sparrows, besides a great deal of butcher's meat.

His plumage very soon began to show itself, crowding off tufts of the down. The quills on his great wings sprouted and grew apace. What a ragged, uncanny appearance he presented! but his look of extreme age gradually became modified. What a lover of the sunlight he was! We would put him out upon the grass in the full blaze of the morning sun, and he would spread his wings and bask in it with the most intense enjoyment. In the nest the young must be exposed to the full power of the midday sun during our first heated terms in June and July, the thermometer often going up to ninety-three or ninety-five degrees, so that sunshine seemed to be a need of [115] his nature. He liked the rain equally well, and when put out in a shower would sit down and take it as if every drop did him good.

His legs developed nearly as slowly as his wings. He could not stand steadily upon them till about ten days before he was ready to fly. The talons were limp and feeble. When we came with food, he would hobble along toward us like the worst kind of a cripple, drooping and moving his wings, and treading upon his legs from the foot back to the elbow, the foot remaining closed and useless. Like a baby learning to stand, he made many trials before he succeeded. He would rise up on his trembling legs only to fall back again.

One day, in the summer-house, I saw him for the first time stand for a moment squarely upon his legs with the feet fully spread beneath them. He looked about him as if the world suddenly wore a new aspect.

His plumage now grew quite rapidly. One red squirrel a day, chopped fine with an axe, was his ration. He began to hold his game with his foot while he tore it. The study was full of his shed down. His dark-brown mottled plumage began to grow beautiful. The wings drooped a little, but gradually he got control of them, and held them in place. [116]

It was now the 20th of July, and the hawk was about five weeks old. In a day or two he was walking or jumping about the grounds. He chose a position under the edge of a Norway spruce, where he would sit for hours dozing, or looking out upon the landscape. When we brought him game, he would advance to meet us with wings slightly lifted, and uttering a shrill cry. Toss him a mouse or sparrow, and he would seize it with one foot and hop off to his cover, where he would bend above it, spread his plumage, look this way and that, uttering all the time the most exultant and satisfied chuckle.

About this time he began to practice striking with his talons, as an Indian boy might begin practicing with his bow and arrow. He would strike at a dry leaf in the grass, or at a fallen apple, or at some imaginary object. He was learning the use of his weapons. His wings also, — he seemed to feel them sprouting from his shoulders. He would lift them straight up and hold them expanded, and they would seem to quiver with excitement. Every hour in the day he would do this. The pressure was beginning to centre there. Then he would strike playfully at a leaf or a bit of wood, and keep his wings lifted.

The next step was to spring into the air and [117] beat his wings. He seemed now to be thinking entirely of his wings. They itched to be put to use.

A day or two later he would leap and fly several feet. A pile of brush ten or twelve feet below the bank was easily reached. Here he would perch in true hawk fashion, to the bewilderment and scandal of all the robins and catbirds in the vicinity. Here he would dart his eye in all directions, turning his head over and glancing up into the sky.

He was now a lovely creature, fully fledged, and as tame as a kitten. But he was not a bit like a kitten in one respect,—he could not bear to have you stroke or even touch his plumage. He had a horror of your hand, as if it would hopelessly defile him. But he would perch upon it, and allow you to carry him about. If a dog or cat appeared, he was ready to give battle instantly. He rushed up to a little dog one day, and struck him with his foot savagely. He was afraid of strangers, and of any unusual object.

The last week in July he began to fly quite freely, and it was necessary to clip one of his wings. As the clipping embraced only the ends of his primaries, he soon overcame the difficulty, and, by carrying his broad, long tail more on that side, flew with considerable ease. He made longer [118] and longer excursions into the surrounding fields and vineyards, and did not always return. On such occasions we would go to find him and fetch him back.

Late one rainy afternoon he flew away into the vineyard, and when, an hour later, I went after him, he could not be found, and we never saw him again. We hoped hunger would soon drive him back, but we have had no clew to him from that day to this.

[119]

THE WINTER WREN

An old hemlock wood at the head waters of the Delaware is a chosen haunt of the winter wren. His voice fills these dim aisles, as if aided by some marvelous sounding-board. Indeed, his song is very strong for so small a bird, and unites in a remarkable degree brilliancy and plaintiveness. I think of a tremulous, vibrating tongue of silver. You may know it is the song of a wren from its gushing, lyrical character; but you must needs look sharp to see the little minstrel, especially while in the act of singing. He is nearly the color of the ground and the leaves; he never ascends the tall trees, but keeps low, flitting from stump to stump and from root to root, dodging in and out of his hiding-places, and watching all intruders with a suspicious eye. He has a very pert, almost comical look. His tail stands more than perpendicular: it points straight toward his head. He is the least ostentatious singer I know of. He does not strike an attitude, and lift up his head in preparation, and, as it were, clear his throat; but sits there on a log and pours out his music, looking straight before him, or even down at the ground. As a songster, he has but few [120] superiors. I do not hear him after the first week in July.

The winter wren is so called because he sometimes braves our northern winters, but it is rarely that one sees him at this season. I think I have seen him only two or three times in winter in my life. The event of one long walk, recently, in February, was seeing one of these birds. As I followed a byroad, beside a little creek in the edge of a wood, my eye caught a glimpse of a small brown bird darting under a stone bridge. I thought to myself no bird but a wren would take refuge under so small a bridge as that. I stepped down upon it and expected to see the bird dart out at the upper end. As it did not appear, I scrutinized the bank of the little run, covered with logs and brush, a few rods farther up.

Presently I saw the wren curtsying and gesticulating beneath an old log. As I approached he disappeared beneath some loose stones in the bank, then came out again and took another peep at me, then fidgeted about for a moment and disappeared again, running in and out of the holes and recesses and beneath the rubbish like a mouse or a chipmunk. The winter wren may always be known by these squatting, bobbing-out-and-in habits.

As I sought a still closer view of him, he flitted [121] stealthily a few yards up the run and disappeared beneath a small plank bridge near a house.

I wondered what he could feed upon at such a time. There was a light skim of snow upon the ground, and the weather was cold. The wren, so far as I know, is entirely an insect-feeder, and where can he find insects in midwinter in our climate? Probably by searching under bridges, under brush-heaps, in holes and cavities in banks where the sun falls warm. In such places he may find dormant spiders and flies and other hibernating insects or their larvæ. We have a tiny, mosquito-like creature that comes forth in March or in midwinter, as soon as the temperature is a little above freezing. One may see them performing their fantastic air-dances when the air is so chilly that one buttons his overcoat about him in his walk. They are darker than the mosquito,—a sort of dark water-color,—and are very frail to the touch. Maybe the wren knows the hiding-place of these insects.

[122]

THE CEDAR-BIRD

How alert and vigilant the birds are, even when absorbed in building their nests! In an open space in the woods I see a pair of cedar-birds collecting moss from the top of a dead tree. Following the direction in which they fly, I soon discover the nest placed in the fork of a small soft maple, which stands amid a thick growth of wild cherry-trees and young beeches. Carefully concealing myself beneath it, without any fear that the workmen will hit me with a chip or let fall a tool, I await the return of the busy pair. Presently I hear the well-known note, and the female sweeps down and settles unsuspectingly into the half-finished structure. Hardly have her wings rested before her eye has penetrated my screen, and with a hurried movement of alarm she darts away. In a moment the male, with a tuft of wool in his beak (for there is a sheep pasture near), joins her, and the two reconnoitre the premises from the surrounding bushes. With their beaks still loaded, they flit round with a frightened look, and refuse to approach the nest till I have moved off and lain down behind a log. Then one of them ventures to alight upon [123] the nest, but, still suspecting all is not right, quickly darts away again. Then they both together come, and after much peeping and spying about, and apparently much anxious consultation, cautiously proceed to work. In less than half an hour it would seem that wool enough has been brought to supply the whole family, real and prospective, with socks, if needles and fingers could be found fine enough to knit it up. In less than a week the female has begun to deposit her eggs,—four of them in as many days,—white tinged with purple, with black spots on the larger end. After two weeks of incubation the young are out.

Excepting the American goldfinch, this bird builds later in the season than any other, its nest, in our northern climate, seldom being undertaken till July. As with the goldfinch, the reason is, probably, that suitable food for the young cannot be had at an earlier period.

I knew a pair of cedar-birds, one season, to build in an apple-tree, the branches of which rubbed against the house. For a day or two before the first straw was laid, I noticed the pair carefully exploring every branch of the tree, the female taking the lead, the male following her with an anxious note and look. It was evident that the wife was to have her choice this time; [124] and, like one who thoroughly knew her mind, she was proceeding to take it. Finally the site was chosen upon a high branch, extending over one low wing of the house. Mutual congratulations and caresses followed, when both birds flew away in quest of building-material. That most freely used is a sort of cotton-bearing plant which grows in old worn-out fields. The nest is large for the size of the bird, and very soft. It is in every respect a first-class domicile.

The cedar-bird is the most silent bird we have. Our neutral-tinted birds, like him, as a rule are our finest songsters; but he has no song or call, uttering only a fine bead-like note on taking flight. This note is the cedar-berry rendered back in sound. When the ox-heart cherries, which he has only recently become acquainted with, have had time to enlarge his pipe and warm his heart, I shall expect more music from him. But in lieu of music, what a pretty compensation are those minute, almost artificial-like, plumes of orange and vermilion that tip the ends of his wing quills! Nature could not give him these and a song too.

[125]

THE GOLDFINCH

About the most noticeable bird of August in New York and New England is the yellowbird, or goldfinch. This is one of the last birds to nest, seldom hatching its eggs till late in July. It seems as if a particular kind of food were required to rear its brood, which cannot be had at an earlier date. The seed of the common thistle is apparently its mainstay. There is no prettier sight at this season than a troop of young goldfinches, led by their parents, going from thistle to thistle along the roadside and pulling the ripe heads to pieces for the seed. The plaintive call of the young is one of the characteristic August sounds. Their nests are frequently destroyed, or the eggs thrown from them, by the terrific July thunder-showers. Last season a pair had a nest on the slender branch of a maple in front of the door of the house where I was staying. The eggs were being deposited, and the happy pair had a loving conversation about them many times each day, when one afternoon a very violent storm arose which made the branches of the trees stream out like wildly disheveled hair, quite turning over those on the windward side, and emptying the [126] pretty nest of its eggs. In such cases the birds build anew,—a delay that may bring the incubation into August.

It is a deep, snug, compact nest, with no loose ends hanging, placed in the fork of a small limb of an apple-tree, a peach-tree, or an ornamental shade-tree. The eggs are faint bluish-white.

While the female is sitting, the male feeds her regularly. She calls to him on his approach, or when she hears his voice passing by, in the most affectionate, feminine, childlike tones, the only case I know where the sitting bird makes any sound while in the act of incubation. When a rival male invades the tree, or approaches too near, the male whose nest it holds pursues and reasons or expostulates with him in the same bright, amicable, confiding tones. Indeed, most birds make use of their sweetest notes in war. The song of love is the song of battle too. The male yellowbirds flit about from point to point, apparently assuring each other of the highest sentiments of esteem and consideration, at the same time that one intimates to the other that he is carrying his joke a little too far. It has the effect of saying with mild and good-humored surprise, "Why, my dear sir, this is my territory; you surely do not mean to trespass; permit me to salute you, and to escort you over the line." [127] Yet the intruder does not always take the hint. Occasionally the couple have a brief sparring-match in the air, and mount up and up, beak to beak, to a considerable height, but rarely do they actually come to blows.

The yellowbird becomes active and conspicuous after the other birds have nearly all withdrawn from the stage and become silent, their broods reared and flown. August is his month, his festive season. It is his turn now. The thistles are ripening their seeds, and his nest is undisturbed by jay-bird or crow. He is the first bird I hear in the morning, circling and swinging through the air in that peculiar undulating flight, and calling out on the downward curve of each stroke, "Here we go, here we go!" Every hour in the day he indulges in his circling, billowy flight. It is a part of his musical performance. His course at such times is a deeply undulating line, like the long, gentle roll of the summer sea, the distance from crest to crest or from valley to valley being probably thirty feet; this distance is made with but one brief beating of the wings on the downward curve. As he quickly opens them, they give him a strong upward impulse, and he describes the long arc with them closely folded. Thus, falling and recovering, rising and sinking like dolphins in the sea, he courses through [128] the summer air. In marked contrast to this feat is his manner of flying when he indulges in a brief outburst of song on the wing. Now he flies level, with broad expanded wings nearly as round and as concave as two shells, which beat the air slowly. The song is the chief matter now, and the wings are used only to keep him afloat while delivering it. In the other case, the flight is the main concern, and the voice merely punctuates it.

Among our familiar birds the matchmaking of none other is quite so pretty as that of the goldfinch. The goldfinches stay with us in loose flocks and clad in a dull-olive suit throughout the winter. In May the males begin to put on their bright summer plumage. This is the result of a kind of superficial moulting. Their feathers are not shed, but their dusky covering or overalls are cast off. When the process is only partly completed, the bird has a smutty, unpresentable appearance. But we seldom see them at such times. They seem to retire from society. When the change is complete, and the males have got their bright uniforms of yellow and black, the courting begins. All the goldfinches of a neighborhood collect together and hold a sort of musical festival. To the number of many dozens [129] they may be seen in some large tree, all singing and calling in the most joyous and vivacious manner. The males sing, and the females chirp and call. Whether there is actual competition on a trial of musical abilities of the males before the females or not, I do not know. The best of feeling seems to pervade the company; there is no sign of quarreling or fighting; "all goes merry as a marriage bell," and the matches seem actually to be made during these musical picnics. Before May is passed the birds are seen in couples, and in June housekeeping usually begins. This I call the ideal of love-making among birds, and is in striking contrast to the squabbles and jealousies of most of our songsters.

I have known the goldfinches to keep up this musical and love-making festival through three consecutive days of a cold northeast rainstorm. Bedraggled, but ardent and happy, the birds were not to be dispersed by wind or weather.

[130]

THE HEN-HAWK [1]

August is the month of the high-sailing hawks. The hen-hawk is the most noticeable. He likes the haze and calm of these long, warm days. He is a bird of leisure, and seems always at his ease. How beautiful and majestic are his movements! So self-poised and easy, such an entire absence of haste, such a magnificent amplitude of circles and spirals, such a haughty, imperial grace, and, occasionally, such daring aerial evolutions!

With slow, leisurely movement, rarely vibrating his pinions, he mounts and mounts in an ascending spiral till he appears a mere speck against the summer sky; then, if the mood seizes him, with wings half closed, like a bent bow, he will cleave the air almost perpendicularly, as if intent on dashing himself to pieces against the earth; but on nearing the ground he suddenly mounts again on broad, expanded wing, as if rebounding upon the air, and sails leisurely away. It is the sublimest feat of the season. One holds his breath till he sees him rise again.

If inclined to a more gradual and less precipitous [131] descent, he fixes his eye on some distant point in the earth beneath him, and thither bends his course. He is still almost meteoric in his speed and boldness. You see his path down the heavens, straight as a line; if near, you hear the rush of his wings; his shadow hurtles across the fields, and in an instant you see him quietly perched upon some low tree or decayed stub in a swamp or meadow, with reminiscences of frogs and mice stirring in his maw.

When the south wind blows, it is a study to see three or four of these air-kings at the head of the valley far up toward the mountain, balancing and oscillating upon the strong current; now quite stationary, except for a slight tremulous motion like the poise of a rope-dancer, then rising and falling in long undulations, and seeming to resign themselves passively to the wind; or, again, sailing high and level far above the mountain's peak, no bluster and haste, but, as stated, occasionally a terrible earnestness and speed. Fire at one as he sails overhead, and, unless wounded badly, he will not change his course or gait.

The calmness and dignity of this hawk, when attacked by crows or the kingbird, are well worthy of him. He seldom deigns to notice his noisy and furious antagonists, but deliberately wheels [132] about in that aerial spiral, and mounts and mounts till his pursuers grow dizzy and return to earth again. It is quite original, this mode of getting rid of an unworthy opponent,—rising to heights where the braggart is dazed and bewildered and loses his reckoning! I am not sure but it is worthy of imitation.

[1] The red-tailed and red-shouldered hawks are both called hen-hawks.

[133]

THE RUFFED GROUSE, OR PARTRIDGE

Whir! whir! whir! and a brood of half-grown partridges start up like an explosion, a few paces from me, and, scattering, disappear into the bushes on all sides. Let me sit down here behind the screen of ferns and briers, and hear this wild hen of the woods call together her brood. At what an early age the partridge flies! Nature seems to concentrate her energies on the wing, making the safety of the bird a point to be looked after first; and while the body is covered with down, and no signs of feathers are visible there, the wing-quills sprout and unfold, and in an incredibly short time the young make fair headway in flying.

Hark! there arises over there in the brush a soft, persuasive cooing, a sound so subtle and wild and unobtrusive that it requires the most alert and watchful ear to hear it. How gentle and solicitous and full of yearning love! It is the voice of the mother hen. Presently a faint timid "Yeap!" which almost eludes the ear, is heard in various directions,—the young responding. As no danger seems near, the cooing of the parent bird is soon a very audible clucking call, [134] and the young move cautiously in that direction. Let me step never so carefully from my hiding-place, and all sounds instantly cease, and I search in vain for either parent or young.

The partridge is one of our native and most characteristic birds. The woods seem good to be in where I find him. He gives a habitable air to the forest, and one feels as if the rightful occupant were really at home. The woods where I do not find him seem to want something, as if suffering from some neglect of Nature. And then he is such a splendid success, so hardy and vigorous. I think he enjoys the cold and the snow. His wings seem to rustle with more fervency in midwinter. If the snow falls very fast, and promises a heavy storm, he will complacently sit down and allow himself to be snowed under. When you approach him at such times, he suddenly bursts out of the snow at your feet, scattering the flakes in all directions, and goes humming away through the woods like a bombshell,—a picture of native spirit and success.

His drum is one of the most welcome and beautiful sounds of spring. Scarcely have the trees expanded their buds, when, in the still April mornings, or toward nightfall, you hear the hum of his devoted wings. He selects, not, as you would predict, a dry and resinous log, but a decayed [135] and crumbling one, seeming to give the preference to old oak-logs that are partly blended with the soil. If a log to his taste cannot be found, he sets up his altar on a rock, which becomes resonant beneath his fervent blows. Who has seen the partridge drum? It is the next thing to catching a weasel asleep, though by much caution and tact it may be done. He does not hug the log, but stands very erect, expands his ruff, gives two introductory blows, pauses half a second, and then resumes, striking faster and faster till the sound becomes a continuous, unbroken whir, the whole lasting less than half a minute. The tips of his wings barely brush the log, so that the sound is produced rather by the force of the blows upon the air and upon his own body as in flying. One log will be used for many years, though not by the same drummer. It seems to be a sort of temple and held in great respect. The bird always approaches on foot, and leaves it in the same quiet manner, unless rudely disturbed. He is very cunning, though his wit is not profound. It is difficult to approach him by stealth; you will try many times before succeeding; but seem to pass by him in a great hurry, making all the noise possible, and with plumage furled he stands as immovable as a knot, allowing you a good view. [136]

The sharp-rayed track of the partridge adds another figure to the fantastic embroidery upon the winter snow. Her course is a clear, strong line, sometimes quite wayward, but generally very direct, steering for the densest, most impenetrable places,—leading you over logs and through brush, alert and expectant, till, suddenly, she bursts up a few yards from you, and goes humming through the trees,—the complete triumph of endurance and vigor. Hardy native bird, may your tracks never be fewer, or your visits to the birch-tree less frequent!

[137]

THE PARTRIDGE

> List the booming from afar,
> Soft as hum of roving bee,
> Vague as when on distant bar

Fall the cataracts of the sea.

Yet again, a sound astray,
Was it the humming of the mill?
Was it cannon leagues away?
Or dynamite beyond the hill?

'T is the grouse with kindled soul,
Wistful of his mate and nest,
Sounding forth his vernal roll
On his love-enkindled breast.

List his fervid morning drum,
List his summons soft and deep,
Calling Spice-bush till she come,
Waking Bloodroot from her sleep.

Ah! ruffled drummer, let thy wing
Beat a march the days will heed,
Wake and spur the tardy spring,
Till minstrel voices jocund ring,
And spring is spring in very deed.

[138]

THE CROW

The crow may not have the sweet voice which the fox in his flattery attributed to him, but he has a good, strong, native speech nevertheless. How much character there is in it! How much thrift and independence! Of course his plumage is firm, his color decided, his wit quick. He understands you at once and tells you so; so does the hawk by his scornful, defiant *whir-r-r-r*. Hardy, happy outlaws, the crows, how I love them! Alert, social, republican, always able to look out for himself, not afraid of the cold and the snow, fishing when flesh is scarce, and stealing when other resources fail, the crow is a character I would not willingly miss from the landscape. I love to see his track in the snow or the mud, and his graceful pedestrianism about the brown fields.

He is no interloper, but has the air and manner of being thoroughly at home, and in rightful possession of the land. He is no sentimentalist like some of the plaining, disconsolate song-birds, but apparently is always in good health and good spirits. No matter who is sick, or dejected, or unsatisfied, or what the weather is, or [139] what the price of corn, the crow is well and finds life sweet. He is the dusky embodiment of worldly wisdom and prudence. Then he is one of Nature's self-appointed constables and greatly magnifies his office. He would fain arrest every hawk or owl or grimalkin that ventures abroad. I have known a posse of them to beset the fox and cry "Thief!" till Reynard hid himself for shame. Do I say the fox flattered the crow when he told him he had a sweet voice? Yet one of the most musical sounds in nature proceeds from the crow. All the crow tribe, from the blue jay up, are capable of certain low ventriloquial notes that have peculiar cadence and charm. I often hear the crow indulging in his in winter, and am reminded of the sound of the dulcimer. The bird stretches up and exerts himself like a cock in the act of crowing, and gives forth a peculiarly clear, vitreous sound that is sure to arrest and reward your attention. This is, no doubt, the song the fox begged to be favored with, as in delivering it the crow must inevitably let drop the piece of meat.

The crow has fine manners. He always has the walk and air of a lord of the soil. One morning I put out some fresh meat upon the snow near my study window. Presently a crow came and carried it off, and alighted with it upon the [140] ground in the vineyard. While he was eating it, another crow came, and, alighting a few yards away, slowly walked up to within a few feet of this fellow and stopped. I expected to see a struggle over the food, as would have been the case with domestic fowls or animals. Nothing of the kind. The feeding crow stopped eating, regarded the other for a moment, made a gesture or two, and flew away. Then the second crow went up to the food, and proceeded to take his share. Presently the first crow came back, when each seized a portion of the food and flew away with it. Their mutual respect and good-will seemed perfect. Whether it really was so in our human sense, or whether it was simply an illustration of the instinct of mutual support which seems to prevail among gregarious birds, I know not. Birds that are solitary in their habits, like hawks or woodpeckers, behave quite differently toward each other in the presence of their food.

The crow will quickly discover anything that looks like a trap or snare set to catch him, but it takes him a long time to see through the simplest contrivance. As I have above stated, I sometimes place meat on the snow in front of my study window to attract him. On one occasion, after a couple of crows had come to expect something there daily, I suspended a piece of meat by a [141] string from a branch of the tree just over the spot where I usually placed the food. A crow soon discovered it, and came into the tree to see what it meant. His suspicions were aroused. There was some design in that suspended meat, evidently. It was a trap to catch him. He surveyed it from every near branch. He peeked and pried, and was bent on penetrating the mystery. He flew to the ground, and walked about and surveyed it from all sides. Then he took a long walk down about the vineyard as if in hope of hitting upon some clew. Then he came to the tree again, and tried first one eye, then the other, upon it; then to the ground beneath; then he went away and came back; then his fellow came, and they both squinted and investigated, and then disappeared. Chickadees and woodpeckers would alight upon the meat and peck it swinging in the wind, but the crows were fearful. Does this show reflection? Perhaps it does, but I look upon it rather as that instinct of fear and cunning so characteristic of the crow. Two days passed thus: every morning the crows came and surveyed the suspended meat from all points in the tree, and then went away. The third day I placed a large bone on the snow beneath the suspended morsel. Presently one of the crows appeared in the tree, and bent his eye upon the [142] tempting bone. "The mystery deepens," he seemed to say to himself. But after half an hour's investigation, and after approaching several times within a few feet of the food upon the ground, he seemed to conclude there was no connection between it and the piece hanging by the string. So he finally walked up to it and fell to pecking it, flickering his wings all the time, as a sign of his watchfulness. He also turned up his eye, momentarily, to the piece in the air above, as if it might be some disguised sword of Damocles ready to fall upon him. Soon his mate came and alighted on a low branch of the tree. The feeding crow regarded him a moment, and then flew up to his side, as if to give him a turn at the meat. But he refused to run the risk. He evidently looked upon the whole thing as a delusion and a snare, and presently went away, and his mate followed him. Then I placed the bone in one of the main forks of the tree, but the crows kept at a safe distance from it. Then I put it back to the ground, but they grew more and more suspicious; some evil intent in it all, they thought. Finally a dog carried off the bone, and the crows ceased to visit the tree.

From my boyhood I have seen the yearly meeting of the crows in September or October, on a [143] high grassy hill or a wooded ridge. Apparently, all the crows from a large area assemble at these times; you may see them coming, singly or in loose bands, from all directions to the rendezvous, till there are hundreds of them together. They make black an acre or two of ground. At intervals they all rise in the air, and wheel about, all cawing at once. Then to the ground again, or to the tree-tops, as the case may be; then, rising again, they send forth the voice of the multitude. What does it all mean? I notice that this rally is always preliminary to their going into winter quarters. It would be interesting to know just the nature of the communication that takes place between them.

[144]

THE CROW

I

My friend and neighbor through the year,
Self-appointed overseer

Of my crops of fruit and grain,
Of my woods and furrowed plain,

Claim thy tithings right and left,
I shall never call it theft.

Nature wisely made the law,
And I fail to find a flaw

In thy title to the earth,
And all it holds of any worth.

I like thy self-complacent air,
I like thy ways so free from care,

Thy landlord stroll about my fields,
Quickly noting what each yields;

Thy courtly mien and bearing bold,
As if thy claim were bought with gold;

Thy floating shape against the sky,
When days are calm and clouds are high;

[145] Thy thrifty flight ere rise of sun,
Thy homing clans when day is done.

Hues protective are not thine,
So sleek thy coat each quill doth shine.

Diamond black to end of toe,
Thy counterpoint the crystal snow.

II

Never plaintive nor appealing,
Quite at home when thou art stealing,

Always groomed to tip of feather,
Calm and trim in every weather,

Morn till night my woods policing,
Every sound thy watch increasing.

Hawk and owl in tree-top hiding
Feel the shame of thy deriding.

Naught escapes thy observation,
None but dread thy accusation.

III

Hunters, prowlers, woodland lovers
Vainly seek the leafy covers.

[146] Noisy, scheming, and predacious,
With demeanor almost gracious,

Dowered with leisure, void of hurry,
Void of fuss and void of worry,

Friendly bandit, Robin Hood,
Judge and jury of the wood,

Or Captain Kidd of sable quill,
Hiding treasures in the hill,

Nature made thee for each season,
Gave thee wit for ample reason,

Good crow wit that's always burnished
Like the coat her care has furnished.

May thy numbers ne'er diminish!
I'll befriend thee till life's finish.

May I never cease to meet thee!
May I never have to eat thee!

And mayest thou never have to fare so
That thou playest the part of scarecrow!

[147]

THE NORTHERN SHRIKE

Usually the character of a bird of prey is well defined; there is no mistaking him. His claws, his beak, his head, his wings, in fact his whole build, point to the fact that he subsists upon live creatures; he is armed to catch them and to slay them. Every bird knows a hawk and knows him from the start, and is on the lookout for him. The hawk takes life, but he does it to maintain his own, and it is a public and universally known fact. Nature has sent him abroad in that character, and has advised all creatures of it. Not so with the shrike; here she has concealed the character of a murderer under a form as innocent as that of the robin. Feet, wings, tail, color, head, and general form and size are all those of a song-bird,—very much like that master songster, the mockingbird,—yet this bird is a regular Bluebeard among its kind. Its only characteristic feature is its beak, the upper mandible having two sharp processes and a sharp hooked point. It usually impales its victim upon a thorn, or thrusts it in the fork of a limb. For the most part, however, its food seems to consist of insects,—spiders, grasshoppers, beetles, etc. It is the assassin of the small birds, whom it often destroys in pure wantonness, or merely to sup on their brains, as the Gaucho slaughters a wild cow or bull for its tongue. It is a wolf in sheep's clothing. Apparently its victims are unacquainted with its true character and allow it to approach them, when the fatal blow is given. I saw an illustration of this the other day. A large number of goldfinches in their fall plumage, together with snowbirds and sparrows, were feeding and chattering in some low bushes back of the barn. I had paused by the fence and was peeping through at them, hoping to get a glimpse of that rare sparrow, the white-crowned. Presently I heard a rustling among the dry leaves as if some larger bird were also among them. Then I heard one of the goldfinches cry out as if in distress, when the whole flock of them started up in alarm, and, circling around, settled in the tops of the larger trees. I continued my scrutiny of the bushes, when I saw a large bird, with some object in its beak, hopping along on a low branch near the ground. It disappeared from my sight for a few moments, then came up through the undergrowth into the top of a young maple where some of the finches had alighted, and I beheld the shrike. The little birds avoided him and flew about the tree, their pursuer following them with the motions [149] of his head and body as if he would fain arrest them by his murderous gaze. The birds did not utter the cry or make the demonstration of alarm they usually do on the appearance of a hawk, but chirruped and called and flew about in a half wondering, half bewildered manner. As they flew farther along the line of trees the shrike followed them as if bent on further captures. I then made my way around to see what the shrike had caught, and what he had done with his prey. As I approached the bushes I saw the shrike hastening back. I read his intentions at once. Seeing my movements, he had returned for his game. But I was too quick for him, and he got up out of the brush and flew away from the locality. On some twigs in the thickest part of the bushes I found his victim,—a goldfinch. It was not impaled upon a thorn, but was carefully disposed upon some horizontal twigs,—laid upon the shelf, so to speak. It was as warm as in life, and its plumage was unruffled. On examining it I found a large bruise or break in the skin on the back of the neck, at the base of the skull. Here the bandit had no doubt gripped the bird with his strong beak. The shrike's bloodthirstiness was seen in the fact that he did not stop to devour his prey, but went in quest of more, as if opening a market of goldfinches. The [150] thicket was his shambles, and if not interrupted, he might have had a fine display of titbits in a short time.

The shrike is called a butcher from his habit of sticking his meat upon hooks and points; further than that, he is a butcher because he devours but a trifle of what he slays.

[151]

THE SCREECH OWL

At one point in the grayest, most shaggy part of the woods, I come suddenly upon a brood of screech owls, full grown, sitting together upon a dry, moss-draped limb, but a few feet from the ground. I pause within four or five yards of them and am looking about me, when my eye lights upon these gray, motionless figures. They sit perfectly upright, some with their backs and some with their breasts toward me, but every head turned squarely in my direction. Their eyes are closed to a mere black line; through this crack they are watching me, evidently thinking themselves unobserved. The spectacle is weird and grotesque, and suggests something impish and uncanny. It is a new effect, the night side of the woods by daylight. After observing them a moment I take a single step toward them, when, quick as thought, their eyes fly wide open, their attitude is changed, they bend, some this way, some that, and, instinct with life and motion, stare wildly around them. Another step, and they all take flight but one, which stoops low on the branch, and with the look of a frightened cat regards me for a few seconds over its shoulder. [152] They fly swiftly and softly, and disperse through the trees.

A winter neighbor of mine, in whom I am interested, and who perhaps lends me his support after his kind, is a little red owl, whose retreat is in the heart of an old apple-tree just over the fence. Where he keeps himself in spring and summer, I do not know, but late every fall, and at intervals all winter, his hiding-place is discovered by the jays and nuthatches, and proclaimed from the tree-tops for the space of half an hour or so, with all the powers of voice they can command. Four times during one winter they called me out to behold this little ogre feigning sleep in his den, sometimes in one apple-tree, sometimes in another. Whenever I heard their cries, I knew my neighbor was being berated. The birds would take turns at looking in upon him, and uttering their alarm-notes. Every jay within hearing would come to the spot, and at once approach the hole in the trunk or limb, and with a kind of breathless eagerness and excitement take a peep at the owl, and then join the outcry. When I approached they would hastily take a final look, and then withdraw and regard my movements intently. After accustoming my eye to the faint light of the cavity for a few [153] moments, I could usually make out the owl at the bottom feigning sleep. Feigning, I say, because this is what he really did, as I first discovered one day when I cut into his retreat with the axe. The loud blows and the falling chips did not disturb him at all. When I reached in a stick and pulled him over on his side, leaving one of his wings spread out, he made no attempt to recover himself, but lay among the chips and fragments of decayed wood, like a part of themselves. Indeed, it took a sharp eye to distinguish him. Not till I had pulled him forth by one wing, rather rudely, did he abandon his trick of simulated sleep or death. Then, like a detected pickpocket, he was suddenly transformed into another creature. His eyes flew wide open, his talons clutched my finger, his ears were depressed, and every motion and look said, "Hands off, at your peril." Finding this game did not work, he soon began to "play possum" again. I put a cover over my study wood-box and kept him captive for a week. Look in upon him at any time, night or day, and he was apparently wrapped in the profoundest slumber; but the live mice which I put into his box from time to time found his sleep was easily broken; there would be a sudden rustle in the box, a faint squeak, and then silence. After a week of captivity I gave him his freedom [154] in the full sunshine; no trouble for him to see which way and where to go.

Just at dusk in the winter nights, I often hear his soft *bur-r-r-r*, very pleasing and bell-like. What a furtive, woody sound it is in the winter stillness, so unlike the harsh scream of the hawk! But all the ways of the owl are ways of softness and duskiness. His wings are shod with silence, his plumage is edged with down.

Another owl neighbor of mine, with whom I pass the time of day more frequently than with the last, lives farther away. I pass his castle every night on my way to the post-office, and in winter, if the hour is late enough, am pretty sure to see him standing in his doorway, surveying the passers-by and the landscape through narrow slits in his eyes. For four successive winters now have I observed him. As the twilight begins to deepen, he rises up out of his cavity in the apple-tree, scarcely faster than the moon rises from behind the hill, and sits in the opening, completely framed by its outlines of gray bark and dead wood, and by his protective coloring virtually invisible to every eye that does not know he is there. Probably my own is the only eye that has ever penetrated his secret, and mine never would have done so had I not chanced on one occasion to see him leave his retreat [155] and make a raid upon a shrike that was impaling a shrew-mouse upon a thorn in a neighboring tree, and which I was watching. I was first advised of the owl's presence by seeing him approaching swiftly on silent, level wing. The shrike did not see him till the owl was almost within the branches. He then dropped his game, and darted back into the thick cover, uttering a loud, discordant squawk, as one would say, "Scat! scat! scat!" The owl alighted, and was, perhaps, looking about him for the shrike's impaled game, when I drew near. On seeing me, he reversed his movement precipitately, flew straight back to the old tree, and alighted in the entrance to the cavity. As I approached, he did not so much seem to move as to diminish in size, like an object dwindling in the distance; he depressed his plumage, and, with his eye fixed upon me, began slowly to back and sidle into his retreat till he faded from my sight. The shrike wiped his beak upon the branches, cast an eye down at me and at his lost mouse, and then flew away.

A few nights afterward, as I passed that way, I saw the little owl again sitting in his doorway, waiting for the twilight to deepen, and undisturbed by the passers-by; but when I paused to observe him, he saw that he was discovered, and he slunk back into his den as on the former occasion. [156] Ever since, while going that way, I have been on the lookout for him. Dozens of teams and foot-passengers pass him late in the day, but he regards them not, nor they him. When I come along and pause to salute him, he opens his eyes a little wider, and, appearing to recognize me, quickly shrinks and fades into the background of his door in a very weird and curious manner. When he is not at his outlook, or when he is, it requires the best powers of the eye to decide the point, as the empty cavity itself is almost an exact image of him. If the whole thing had been carefully studied, it could not have answered its purpose better. The owl stands quite perpendicular, presenting a front of light mottled gray; the eyes are closed to a mere slit, the ear-feathers depressed, the beak buried in the plumage, and the whole attitude is one of silent, motionless waiting and observation. If a mouse should be seen crossing the highway, or scudding over any exposed part of the snowy surface in the twilight, the owl would doubtless swoop down upon it. I think the owl has learned to distinguish me from the rest of the passers-by; at least, when I stop before him, and he sees himself observed, he backs down into his den, as I have said, in a very amusing manner.

[157]

THE CHICKADEE

The chickadees we have always with us. They are like the evergreens among trees and plants. Winter has no terrors for them. They are properly wood-birds, but the groves and orchards know them also. Did they come near my cabin for better protection, or did they chance to find a little cavity in a tree there that suited them? Branch-builders and ground-builders are easily accommodated, but the chickadee must find a cavity, and a small one at that. The woodpeckers make a cavity when a suitable trunk or branch is found, but the chickadee, with its small, sharp beak, rarely does so; it usually smooths and deepens one already formed. This a pair did a few yards from my cabin. The opening was into the heart of a little sassafras, about four feet from the ground. Day after day the birds took turns in deepening and enlarging the cavity: a soft, gentle hammering for a few moments in the heart of the little tree, and then the appearance of the worker at the opening, with the chips in his, or her, beak. They changed off every little while, one working while the other gathered food. Absolute equality of the sexes, [158] both in plumage and in duties, seems to prevail among these birds, as among a few other species. During the preparations for housekeeping the birds were hourly seen and heard, but as soon as the first egg was laid, all this was changed. They suddenly became very shy and quiet. Had it not been for the new egg that was added each day, one would have concluded that they had abandoned the place. There was a precious secret now that must be well kept. After incubation began, it was only by watching that I could get a glimpse of one of the birds as it came quickly to feed or to relieve the other.

One day a lot of Vassar girls came to visit me, and I led them out to the little sassafras to see the chickadee's nest. The sitting bird kept her place as head after head, with its nodding plumes and millinery, appeared above the opening to her chamber, and a pair of inquisitive eyes peered down upon her. But I saw that she was getting ready to play her little trick to frighten them away. Presently I heard a faint explosion at the bottom of the cavity, when the peeping girl jerked her head quickly back, with the exclamation, "Why, it spit at me!" The trick of the bird on such occasions is apparently to draw in its breath till its form perceptibly swells, and then give forth a quick, explosive sound like an [159] escaping jet of steam. One involuntarily closes his eyes and jerks back his head. The girls, to their great amusement, provoked the bird into this pretty outburst of her impatience two or three times. But as the ruse failed of its effect, the bird did not keep it up, but let the laughing faces gaze till they were satisfied.

I was much interested in seeing a brood of chickadees, reared on my premises, venture upon their first flight. Their heads had been seen at the door of their dwelling—a cavity in the limb of a pear-tree—at intervals for two or three days. Evidently they liked the looks of the great outside world; and one evening, just before sundown, one of them came forth. His first flight was of several yards, to a locust, where he alighted upon an inner branch, and after some chirping and calling proceeded to arrange his plumage and compose himself for the night. I watched him till it was nearly dark. He did not appear at all afraid there alone in the tree, but put his head under his wing and settled down for the night as if it were just what he had always been doing. There was a heavy shower a few hours later, but in the morning he was there upon his perch in good spirits.

I happened to be passing in the morning when another one came out. He hopped out upon [160] a limb, shook himself, and chirped and called loudly. After some moments an idea seemed to strike him. His attitude changed, his form straightened up, and a thrill of excitement seemed to run through him. I knew what it all meant; something had whispered to the bird, "Fly!" With a spring and a cry he was in the air, and made good headway to a near hemlock. Others left in a similar manner during that day and the next, till all were out.

[161]

THE DOWNY WOODPECKER

The bird that seems to consider he has the best right to my hospitality is the downy woodpecker, my favorite neighbor among the winter birds. His retreat is but a few paces from my own, in the decayed limb of an apple-tree, which he excavated several autumns ago. I say "he" because the red plume on the top of his head proclaims the sex. It seems not to be generally known to our writers upon ornithology that certain of our woodpeckers—probably all the winter residents—each fall excavate a limb or the trunk of a tree in which to pass the winter, and that the cavity is abandoned in the spring, probably for a new one in which nidification takes place.

DOWNY WOODPECKER

The particular woodpecker to which I refer drilled his first hole in my apple-tree one fall four or five years ago. This he occupied till the following spring, when he abandoned it. The next fall he began a hole in an adjoining limb, later than before, and when it was about half completed a female took possession of his old quarters. I am sorry to say that this seemed to enrage the male very much, and he persecuted the poor bird whenever she appeared upon the scene. He [162] would fly at her spitefully and drive her off. One chilly November morning, as I passed under the tree, I heard the hammer of the little architect in his cavity, and at the same time saw the persecuted female sitting at the entrance of the other hole as if she would fain come out. She was actually shivering, probably from both fear and cold. I understood the situation at a glance; the bird was afraid to come forth and brave the anger of the male. Not till I had rapped smartly upon the limb with my stick did she come out and attempt to escape; but she had not gone ten feet from the tree before the male was in hot pursuit, and in a few moments had driven her back to the same tree, where she tried to avoid him among the branches. There is probably no gallantry among the birds except at the mating season. I have frequently seen the male woodpecker drive the female away from the bone upon the tree. When she hopped around to the other end and timidly nibbled it, he would presently dart spitefully at her. She would then take up her position in his rear and wait till he had finished his meal. The position of the female among the birds is very much the same as that of women among savage tribes. Most of the drudgery of life falls upon her, and the leavings of the males are often her lot.

[163] My bird is a genuine little savage, doubtless, but I value him as a neighbor. It is a satisfaction during the cold or stormy winter nights to know he is warm and cozy there in his retreat. When the day is bad and unfit to be abroad in, he is there too. When I wish to know if he is at home, I go and rap upon his tree, and, if he is not too lazy or indifferent, after some delay he shows his head in his round doorway about ten feet above, and looks down inquiringly upon me—sometimes latterly I think half resentfully, as much as to say, "I would thank you not to disturb me so often." After sundown, he will not put his head out any more when I call, but as I step away I can get a glimpse of him inside looking cold and reserved. He is a late riser, especially if it is a cold or disagreeable morning, in this respect being like the barn fowls; it is sometimes near nine o'clock before I see him leave his tree. On the other hand, he comes home early, being in, if the day is unpleasant, by four P.M. He lives all

alone; in this respect I do not commend his example. Where his mate is, I should like to know.

I have discovered several other woodpeckers in adjoining orchards, each of which has a like home, and leads a like solitary life. One of them has excavated a dry limb within easy reach of [164] my hand, doing the work also in September. But the choice of tree was not a good one; the limb was too much decayed, and the workman had made the cavity too large; a chip had come out, making a hole in the outer wall. Then he went a few inches down the limb and began again, and excavated a large, commodious chamber, but had again come too near the surface; scarcely more than the bark protected him in one place, and the limb was very much weakened. Then he made another attempt still farther down the limb, and drilled in an inch or two, but seemed to change his mind; the work stopped, and I concluded the bird had wisely abandoned the tree. Passing there one cold, rainy November day, I thrust in my two fingers and was surprised to feel something soft and warm: as I drew away my hand the bird came out, apparently no more surprised than I was. It had decided, then, to make its home in the old limb; a decision it had occasion to regret, for not long after, on a stormy night, the branch gave way and fell to the ground: —

> "When the bough breaks the cradle will fall,
> And down will come baby and cradle and all."

Another trait our woodpeckers have that endears them to me is their habit of drumming in [165] the spring. They are songless birds, and yet all are musicians; they make the dry limbs eloquent of the coming change. Did you think that loud, sonorous hammering which proceeded from the orchard or from the near woods on that still March or April morning was only some bird getting its breakfast? It is Downy, but he is not rapping at the door of a grub; he is rapping at the door of spring, and the dry limb thrills beneath the ardor of his blows.

A few seasons ago, a downy woodpecker, probably the individual one who is now my winter neighbor, began to drum early in

March in a partly decayed apple-tree that stands in the edge of a narrow strip of woodland near me. When the morning was still and mild I would often hear him through my window before I was up, or by half-past six o'clock, and he would keep it up pretty briskly till nine or ten o'clock, in this respect resembling the grouse, which do most of their drumming in the forenoon. His drum was the stub of a dry limb about the size of one's wrist. The heart was decayed and gone, but the outer shell was hard and resonant. The bird would keep his position there for an hour at a time. Between his drummings he would preen his plumage and listen as if for the response of the female, or for the drum of some rival. How [166] swiftly his head would go when he was delivering his blows upon the limb! His beak wore the surface perceptibly. When he wished to change the key, which was quite often, he would shift his position an inch or two to a knot which gave out a higher, shriller note. When I climbed up to examine his drum, he was much disturbed. I did not know he was in the vicinity, but it seems he saw me from a near tree, and came in haste to the neighboring branches, and with spread plumage and a sharp note demanded plainly enough what my business was with his drum. I was invading his privacy, desecrating his shrine, and the bird was much put out. After some weeks the female appeared; he had literally drummed up a mate; his urgent and oft-repeated advertisement was answered. Still the drumming did not cease, but was quite as fervent as before. If a mate could be won by drumming, she could be kept and entertained by more drumming; courtship should not end with marriage. If the bird felt musical before, of course he felt much more so now. Besides that, the gentle deities needed propitiating in behalf of the nest and young as well as in behalf of the mate. After a time a second female came, when there was war between the two. I did not see them come to blows, but I saw one female pursuing the other about the place, and giving [167] her no rest for several days. She was evidently trying to run her out of the neighborhood. Now and then, she, too, would drum briefly, as if sending a triumphant message to her mate.

The woodpeckers do not each have a particular dry limb to which they resort at all times to drum, like the one I have described. The woods are full of suitable branches, and they drum more or less here and there as they are in quest of food; yet I am convinced each one has its favorite spot, like the grouse, to which it resorts especially in the morning. The sugar-maker in the maple woods may notice that this sound proceeds from the same tree or trees about his camp with great regularity. A woodpecker in my vicinity has drummed for two seasons on a telegraph-pole, and he makes the wires and glass insulators ring. Another drums on a thin board on the end of a long grape-arbor, and on still mornings can be heard a long distance.

I watch these woodpeckers daily to see if I can solve the mystery as to how they hop up and down the trunks and branches without falling away from them when they let go their hold. They come down a limb or trunk backward by a series of little hops, moving both feet together. [168] If the limb is at an angle to the tree and they are on the under side of it, they do not fall away from it to get a new hold an inch or half-inch farther down. They are held to it as steel to a magnet. Both tail and head are involved in the feat. At the instant of making the hop the head is thrown in and the tail thrown out, but the exact mechanics of it I cannot penetrate. Philosophers do not yet know how a backward-falling cat turns in the air, but turn she does. It may be that the woodpecker never quite relaxes his hold, though to my eye he appears to do so.

[169]

THE DOWNY WOODPECKER

> Downy came and dwelt with me,
> Taught me hermit lore;
> Drilled his cell in oaken tree
> Near my cabin door.
>
> Architect of his own home
> In the forest dim,
> Carving its inverted dome
> In a dozy limb.
>
> Carved it deep and shaped it true
> With his little bill;

Took no thought about the view,
Whether dale or hill.

Shook the chips upon the ground,
Careless who might see.
Hark! his hatchet's muffled sound
Hewing in the tree.

Round his door as compass-mark,
True and smooth his wall;
Just a shadow on the bark
Points you to his hall.

[170] Downy leads a hermit life
All the winter through;
Free his days from jar and strife,
And his cares are few.

Waking up the frozen woods,
Shaking down the snows;
Many trees of many moods
Echo to his blows.

When the storms of winter rage,
Be it night or day,
Then I know my little page
Sleeps the time away.

Downy's stores are in the trees,
Egg and ant and grub;
Juicy tidbits, rich as cheese,
Hid in stump and stub.

Rat-tat-tat his chisel goes,
Cutting out his prey;
Every boring insect knows
When he comes its way.

Always rapping at their doors,
Never welcome he;
All his kind, they vote, are bores,
Whom they dread to see.

[171] Why does Downy live alone
In his snug retreat?
Has he found that near the bone
Is the sweetest meat?

Birdie craved another fate
When the spring had come;
Advertised him for a mate
On his dry-limb drum.

Drummed her up and drew her near,
In the April morn,
Till she owned him for her dear
In his state forlorn.

Now he shirks all family cares,
This I must confess;
Quite absorbed in self affairs
In the season's stress.

We are neighbors well agreed
Of a common lot;
Peace and love our only creed
In this charmèd spot.

[173]

www.ingramcontent.com/pod-product-compliance
Lightning Source LLC
Chambersburg PA
CBHW031426210526
45464CB00005B/2064